Cambridge Primary

Ready to Go Lessons for Maths

Step-by-step
lesson plans for
Cambridge Primary

Stage 5

Caroline Clissold

Series editor: Paul Broadbent

HODDER EDUCATION
AN HACHETTE UK COMPANY

Acknowledgements

Every effort has been made to trace all copyright holders, but if any have been inadvertently overlooked the Publishers will be pleased to make the necessary arrangements at the first opportunity.

Although every effort has been made to ensure that website addresses are correct at time of going to press, Hodder Education cannot be held responsible for the content of any website mentioned in this book. It is sometimes possible to find a relocated web page by typing in the address of the home page for a website in the URL window of your browser. Websites included in this text have not been reviewed as part of the Cambridge endorsement process.

Hachette UK's policy is to use papers that are natural, renewable and recyclable products and made from wood grown in sustainable forests. The logging and manufacturing processes are expected to conform to the environmental regulations of the country of origin.

Orders: please contact Bookpoint Ltd, 130 Milton Park, Abingdon, Oxon OX14 4SB. Telephone: (44) 01235 827720. Fax: (44) 01235 400454. Lines are open 9.00–5.00, Monday to Saturday, with a 24-hour message answering service. Visit our website at www.hoddereducation.com.

© Caroline Clissold 2012
First published in 2012 by
Hodder Education,
An Hachette UK Company
338 Euston Road
London NW1 3BH

Impression number   5 4 3 2 1
Year   2016 2015 2014 2013 2012

Cover illustration by Peter Lubach
Illustrations by Planman Technologies
Typeset in ITC Stone Serif 10/12.5 by Planman Technologies
Printed in Great Britain by CPI Group (UK) Ltd, Croydon, CR0 4YY

A catalogue record for this title is available from the British Library

ISBN: 978 1444 177626

Contents

Introduction

About the series

Ready to Go Lessons for Maths is a series of photocopiable resource books providing creative teaching strategies for primary teachers. These books support the revised Cambridge Primary curriculum frameworks for English, Mathematics and Science at Stages 1–6 (ages 5–11). They have been written by experienced primary teachers to reflect the different teaching approaches recommended in the Cambridge Primary Teacher Guides. The books contain lesson plans and photocopiable support materials, with a wide range of activities and appropriate ideas for assessment and differentiation. As the books are intended for international schools we have taken care to ensure that they are culturally sensitive.

Cambridge Primary

The Cambridge Primary curriculum frameworks show schools how to develop the learners' knowledge, skills and understanding in English, Mathematics and Science. They provide a secure foundation in preparation for the Cambridge Secondary 1 (lower secondary) curriculum. The ideas in this book can also be easily incorporated into existing curriculum frameworks already in your school.

How to use this book

This book covers each of the units of the Scheme of work for Mathematics at Stage 5. It can be worked through systematically (as all the learning objectives are covered), or used to support areas where you feel you need more ideas. It is not prescriptive – it gives ideas and suggestions for you to incorporate into your own teaching as you see fit.

Each step-by-step lesson plan shows you the learning objectives you will cover, the resources you will need and how to deliver the lesson. Each lesson includes a Mental / oral starter activity, Main activities and a Plenary that draws the lesson to a close and recaps the learning objectives. Success criteria are provided in the form of questions to help you assess the learners' level of understanding. The Differentiation section provides support for the less-able learners and extension ideas for the more able.

For each lesson plan there is at least one supporting photocopiable activity page. At the end of each unit there are also suggestions for assessment activities. Answers to activities can be found at www.hoddereducation.com/checkpointextras.

Learning objectives

The *Mathematics Curriculum Framework* provides a set of learning objectives for each stage. At the start of each lesson you need to re-phrase the learning objectives into child-friendly language so that you can share them with the learners at the outset. It sometimes helps to express them as *We are learning to / about …* statements. This really does help the learners to focus on the lesson's outcomes. For example: 'Understand what each digit represents in three-digit numbers and partition into hundreds, tens and units.' (Stage 3) could be introduced to the learners at the start of the lesson as: *We are learning about the value of each of the digits in a 3-digit number.* To avoid unnecessary repetition we have not included such statements at the start of each lesson plan but it is understood that the teacher would do this.

The overview chart on pages 6–7 shows you how the learning objectives are covered in the lessons in this book.

Success criteria

These are the measures that the teacher and, eventually, the learner will be able to use to assess the outcome of the learning that has taken place in each lesson. They are included as a series of questions, which will help you as teacher to assess the learners' understanding of the skills and knowledge covered in the lesson.

Problem-solving skills

Maths teaching is concerned with more than just the learning of mathematical facts. Problem-solving skills are also **essential** and are planned as an ongoing and sequential part of each unit.

The activities in this book will show you how to incorporate the practical nature of problem-solving so that it is part of the teaching process. Problem-solving objectives are worked into every teaching unit, with these skills underpinning all other strands to help the learners understand mathematical relationships and functions. These skills need to be used regularly in familiar and new contexts in order for the learners to become mathematical thinkers who are capable of questioning, reasoning and seeking answers through investigation.

The key to successful mathematical problem-solving teaching lies in providing the learners with opportunities to learn by doing, that is, through **active learning**.

Mental / oral starters

These are an important part of each lesson, consisting of whole-class, teacher-led interactive activities. The purpose varies for each lesson, and can include:

- practice and consolidation of existing skills – often mental calculation but also properties of shape and the language of number
- quick recall – to secure knowledge of number facts and build up speed and accuracy
- revisiting previous learning – to return to aspects of Maths that may have caused difficulty or to strengthen the learners' knowledge and use of mental or written strategies
- preparation for the main part of the lesson – linked to the objectives for the lesson to support the learning.

Formative assessment

Formative assessment is on-going assessment that occurs in every lesson and informs the teacher and learners of the progress they are making, linked to the success criteria. The types of questions to ask that will support teachers in making formative assessments have been incorporated into each lesson in the 'Success criteria' sections.

One of the advantages of formative assessment is that any problems that arise during the lesson can be responded to immediately. Formative assessment influences the next steps in learning and may influence changes in planning and / or delivery for subsequent lessons.

Summative assessment

Summative assessment is essential at the end of each unit of work to assess exactly what the learners know, understand and can do. The assessment sections at the end of each unit are designed to provide you with a variety of opportunities to check the learners' understanding of the unit. These activities can include specific questions for teachers to ask, activities for the learners to carry out (independently, in pairs or in groups) or written assessment.

The information gained from both the formative and summative assessment ideas can then be used to inform future planning in order to close any gaps in the learners' understanding as recommended by *Assessment for Learning*.

Appropriate use of ICT

At the planning stage teachers need to consider how the use of ICT in a lesson will enhance the learning process. Ensure that the ICT resources you use support and promote the learners' understanding of the learning objectives. Activities included in this book have been designed to be carried out without the need for state-of-the-art ICT facilities. Suggestions have also been included for schools with internet access and / or the use of interactive whiteboards. This is in order to cater for most teachers' needs.

In these lessons the author sometimes asks for the teacher to display an enlarged version of the photocopiable page at the front of the class. We have not specified whether this should be using an overhead projector, interactive whiteboard or flipchart, as schools will have different resources available to them.

We hope that using these resources will give you confidence and creative ideas in delivering the Cambridge Primary curriculum framework.

**Paul Broadbent,
Series Editor**

Overview chart

Term 1	Unit 1A: Number and problem solving	Lesson	Framework code	Page
		Numbers to 1 000 000	5Nn1 5Nn2 5Nn3	8
		Rounding	5Nn6 5Pt6 5Ps3	10
		Comparing and ordering	5Nc6 5Nn8 5Ps3	12
		Mental calculation strategies 1	5Nn12 5Nc4 5Ps8	14
		Mental calculation strategies 2	5Nc8 5Pt2 5Ps4	16
		Addition and subtraction 1	5Nc10 5Nc18 5Pt2	18
		Addition and subtraction 2	5Nc18 5Pt3 5Ps10	20
		Multiples and factors	5Nc5 5Nc7 5Ps5	22
		Multiplication 1	5Nc3 5Nc20 5Ps9	24
		Multiplication 2	5Nc20 5Nc21 5Pt7	26
		Division 1	5Nn5 5Nc23 5Nc26	28
		Division 2	5Nn13 5Nc25 5Ps2	30
	Unit assessment			32
	Unit 1B: Geometry and problem solving	Triangles	5Gs1 5Ps4	34
		Nets of 3D shapes	5Gs4 5Ps4	36
		Co-ordinates 1	5Gp1 5Ps4 5Ps9	38
		Co-ordinates 2	5Gp1 5Ps4 5Ps9	40
		Transformations 1	5Gp2 5Ps4	42
		Transformations 2	5Gs2 5Ps4 5Ps9	44
		Transformations 3	5Gp3 5Ps4	46
	Unit assessment			48
	Unit 1C: Measures and problem solving	Measures	5Ml2 5Ml3 5Ml4 5Ps9	50
		Length	5Ml1 5Ml2 5Ml7	52
		Mass	5Ml1 5Ml5 5Ps10	54
		Capacity	5Ml1 5Ml2 5Ml6	56
		Time 1	5Mt1 5Mt2 5Pt1	58
		Time 2	5Mt3 5Mt5 5Ps2 5Ps5	60
		Perimeter and area 1	5Ma1 5Ma2 5Ps4 5Pt7	62
		Perimeter and area 2	5Ma1 5Ma2 5Ma3 5Ps4	64
	Unit assessment			66

Term 2	Unit 2A: Number and problem solving	Lesson	Framework code	Page
		Numbers to 1 000 000	5Nn1 5Nn3	68
		Number sequences	5Nc1 5Nc2 5Nn12 5Ps3	70
		10, 100 and 1000	5Nc3 5Nn5	72
		Comparing and ordering	5Nn9 5Ps8	74
		Decimal fractions 1	5Nn4 5Nn7 5Pt7	76
		Decimal fractions 2	5Nn7 5Nn11 5Ps9	78
		Odd and even numbers	5Nn14 5Ps4 5Ps5	80
		Multiples and factors	5Nc4 5Nc5 5Nc7 5Ps3	82
		Mental calculation strategies	5Ps6 5Nc8 5Nc9 5Nc11	84
		Addition and subtraction 1	5Pt6 5Pt3 5Ps10	86
		Addition and subtraction 2	5Nc10 5Nc19 5Pt2 5Ps2	88
		Multiplication 1	5Nc14 5Nc20 5Pt2	90
		Multiplication 2	5Nc12 5Nc13 5Nc15 5Nc21	92
		Division	5Nc16 5Nc17 5Nc23 5Pt4	94
	Unit assessment			96

7

Numbers to 1 000 000

- Count on and back in steps of constant size, extending beyond zero. (5Nn1)
- Know what each digit represents in five- and six-digit numbers. (5Nn2)
- Partition any number up to one million into thousands, hundreds, tens and units. (5Nn3)

Resources

Counting stick; place value grid; pencils; paper; one set of 0 to 9 digit cards from photocopiable page 9 for each learner; partitioning cards.

Starter

- Explain that numbers are the focus of the lesson, so the learners will practise counting in steps of different sizes.
- Show the counting stick. Explain: *One end is zero, the other is 60.* Ask: *What steps should we count in to get from zero to 60?* Agree six. Together, count from zero to 60 and back. Place your finger on different divisions. *What number would be found on these?* Put your finger on the fifth interval. *This is zero.* Together, count in steps of six from zero to one end (30). Then count back to the other (−30).
- Repeat this for steps of seven, eight and nine.

Main activities

- Show the place value grid. Point to the digits in a row in turn, beginning with the highest. Ask the learners to read the whole number. They write it down, for example 400 + 30 + 7 = 437, and say it, for example 'four hundred and thirty-seven'.
- Repeat for four-, five- and six-digit numbers.
- Give the learners a set of digit cards from photocopiable page 9. Ask them to make different four-, five- and six-digit numbers. For each, ask what place value different digits have; for example, in 13 298, the 1 is 10 000.

- In pairs, ask the learners to pick six digit cards, make a number, read it to each other and then write down what each digit represents. They swap two digit cards, then decide whether the new number is higher or lower. They estimate by how much. For example: for 452 871, swapping the 4 and 8 makes a new number approximately 400 000 bigger.

Plenary

- Discuss what they have been learning about during the lesson.
- Write a few six-digit numbers on the board. Point to some of the digits and ask individuals to tell you what they represent.
- Ask where they might see amounts of this size.

Success criteria

Ask the learners:

- What can you tell me about the number 256 498? What else?
- What is each digit in 345 204 worth? What do we call the zero? Why?
- What do we mean by place value? Can you describe this in another way?
- Where might we see 456 390 items? Where else?

Ideas for differentiation

Support: Ask these learners to focus on three- and four-digit numbers. Provide them with partitioning cards, so that they can physically build the numbers and clearly see that, for example, 4563 = 4000 + 500 + 60 + 3.

Extension: Ask these learners to work with millions numbers.

Digit cards

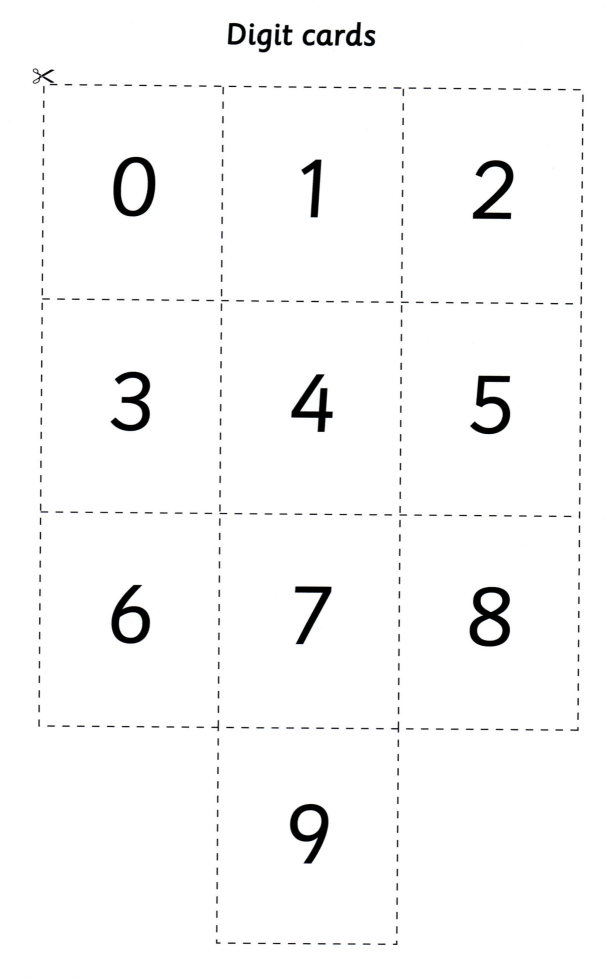

Rounding

Learning objectives

- Round four-digit numbers to the nearest 10, 100 or 1000. (5Nn6)
- Estimate and approximate when calculating, e.g. using rounding, and check working. (5Pt6)
- Explore and solve number problems and puzzles, e.g. logic problems. (5Ps3)

Resources

Set of follow-me cards cut from photocopiable page 11; one set of 0 to 9 digit cards from photocopiable page 9 for each learner; pencils; paper.

Starter

- Tell the learners that they will be revising their times tables.
- Divide the learners into small groups and give some of the follow-me cards from photocopiable page 11 to each group. Keep one card. Play the game as follows, timing how long it takes. Read out the multiplication on your card, for example: *4 × 6*. The group with the answer calls it out. They then read the multiplication on their card. Continue. You will have the last answer. Repeat the activity: are they faster the second time?
- Start with the answer on your card. The group that has the multiplication question that goes with it reads it out. They then read out the next answer, and so on.

Main activities

- Set this problem: *Sam would like to buy a laptop for $589 and a printer for $218. Approximately how much money does he need?* Ask: *What is meant by 'approximately'?* Establish that it is an amount close to the answer but not exact. Ask the learners to discuss in pairs how to find the approximate amount.
- Agree that they can round the amounts.

- Practise rounding. Begin with two-digit numbers and progress to three-, four-, five- and six-digit numbers. Ask them to round to the nearest 10, 100 and 1000 as appropriate; for example 34 583 rounds to 34 580, 34 600, 35 000.
- Return to the problem. Ask the learners to round the amounts to the nearest 10 or 100 and add ($590 + $220 = $810 or $600 + $200 = $800). Set similar problems.
- Ask the learners to use digit cards from photocopiable page 9 to make numbers and then round them to 10, 100 and 1000 as appropriate.

Plenary

- Call out a selection of numbers to round as in the lesson. Ask the learners to write their answers. Call a variety so that all the learners can participate at their own level.
- Discuss when rounding is useful in real life.

Success criteria

Ask the learners:

- What does it mean to round? Can you explain in a different way?
- Why is rounding useful? Can you think of another reason?
- What would be the approximate answer to this calculation: 3467 + 2912? How did you come to that solution? Are there any other answers?

Ideas for differentiation

Support: Ask these learners to round two- and three-digit numbers to the nearest 10 and 100. They should draw simple number lines and plot the numbers to be rounded onto them.

Extension: Ask these learners to work with six-digit numbers and round to the nearest 10, 100, 1000 and 10 000.

Follow-me cards

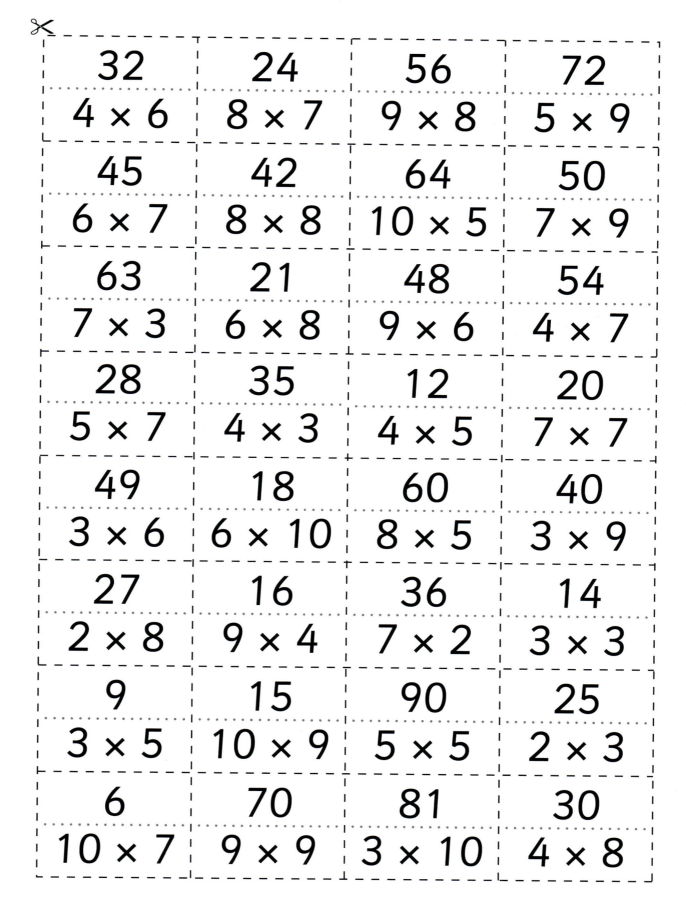

32	24	56	72
4 × 6	8 × 7	9 × 8	5 × 9
45	42	64	50
6 × 7	8 × 8	10 × 5	7 × 9
63	21	48	54
7 × 3	6 × 8	9 × 6	4 × 7
28	35	12	20
5 × 7	4 × 3	4 × 5	7 × 7
49	18	60	40
3 × 6	6 × 10	8 × 5	3 × 9
27	16	36	14
2 × 8	9 × 4	7 × 2	3 × 3
9	15	90	25
3 × 5	10 × 9	5 × 5	2 × 3
6	70	81	30
10 × 7	9 × 9	3 × 10	4 × 8

Comparing and ordering

Learning objectives

- Know squares of all numbers to 10 × 10. (5Nc6)
- Order and compare numbers up to a million using the > and < signs. (5Nn8)
- Explore and solve number problems and puzzles, e.g. logic problems. (5Ps3)

Resources

Pencils; paper; one set of 0 to 9 digit cards from photocopiable page 9 for each learner; photocopiable page 13.

Starter

- Explain that the learners will practise multiples and square numbers. This will help in today's lesson.
- Call out multiples of 6, 7, 8 and 9; for example: *40, 72*. Ask the learners to write down what numbers these are multiples of. Some will be multiples of several. Ask them to find all of them; for example 40: 2, 4, 5, 8, 10, 20.
- Give them one minute to write down as many multiples of 6 as they can. Repeat for 7, 8 and 9.
- Call out squares of different numbers; for example: *81, 49*. The learners write the numbers that need to be squared to give these results.

Main activities

- Write this on the board: $6^2 > \square$. Ask the learners to discuss what this means in pairs. Take feedback. Agree that > is the greater than symbol.
- Ask them to find numbers that could go in the empty box and write them down. Take feedback. Agree that it could be any number less than 36.
- Repeat with other square numbers. Use both > and < symbols.

- Give this statement: $5 \times 8 < \frac{1}{2}$ *of 80*. Ask the learners to work in pairs to find out if this is true or not. Repeat with similar statements.
- Write this on the board: $8^2 < ? \times 6$. Ask the learners to work in pairs to find numbers that can replace the question mark. Repeat with other missing-number sentences.
- Distribute the digit cards from photocopiable page 9 and copies of photocopiable page 13 for the learners to work through.

Plenary

- *What have you been learning about in today's lesson?* Take feedback.
- Ask the learners to write some number sentences using the > or < symbols. Encourage them to be creative.
- Share creative sentences.

Success criteria

Ask the learners:

- How could you find the missing number in this number sentence: $5 \times 8 < ? \times 6$? Is there another way?
- What about this number sentence: $10^2 > ? \times 20$? Is there another answer?
- How might these symbols be useful in everyday life?

Ideas for differentiation

Support: During the lesson these learners should make number sentences using multiples of 5 and 10.

Extension: During the lesson these learners work with five or six digits. In the plenary challenge them to make multiplication and division number sentences with = as well as > and <, for example, $4^2 \times 5 > 6^2 \div 3$, $7^2 \times 2 = 96 + 2$.

Name: _____

Greater or less?

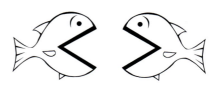

You will need:

A set of 0 to 9 digit cards.

What to do

● Shuffle your digit cards. Place them face down on the table in front of you.

● Pick two cards and multiply them together. Then pick two more cards and multiply them together.

● Write your calculations in the table. Make sure they make a correct number sentence.

● An example has been done for you.

Multiplication	Answer	Greater or less	Multiplication	Answer
6 × 7	42	>	5 × 2	10
		<		
		<		
		>		
		>		
		<		
		>		
		<		
		<		
		>		

● If you found this easy, pick three or four digit cards and multiply them together! Record your new number sentences below:

_____

_____

_____

Mental calculation strategies 1

Learning objectives

● Recognise and extend number sequences. (5Nn12)

● Know and apply tests of divisibility by 2, 5, 10 and 100. (5Nc4)

● Investigate a simple general statement by finding examples which do or do not satisfy it, e.g. the sum of three consecutive whole numbers is always a multiple of three. (5Ps8)

Resources

Set of cards from photocopiable page 11; pencils; paper; photocopiable page 15; 0 to 9 digit cards from photocopiable page 9.

Starter

- Tell the learners that they will be practising their times tables as these will be needed during the lesson.
- Use the cards from photocopiable page 11 to play the follow-me game (as described on page 10) to practise multiplication tables.
- Call out some multiplications; for example: 8×6, 7×9. Ask the learners to write down the answers as quickly as they can.

Main activities

- Write this number sequence on the board: *5, 10, 15, 20.* Ask the learners to continue it to 50. Discuss what the numbers have in common. Establish that they are all in the 5 times table and therefore must all be multiples of 5. Agree that multiples of 5 end with a 5 or 0.
- Repeat for multiples of 2 with this sequence: *10, 12, 14, 16, 18.* Agree that multiples of 2 are even.
- Repeat for multiples of 10 (50, 60, 70, 80) and 100 (500, 700, 800). Agree that multiples of 10 end with zero and multiples of 100 end with two zeros.

- Ask the learners to use these rules to tell you what 20 can be divided by. Agree 2, 5 and 10 because the number is even and ends with zero. Link with the inverse operations of multiplication and division.
- Do this for other numbers, for example 100, 95, 64, 1200.
- Ask the learners to work with a partner and write down six numbers that are multiples of 2, 5, 10 and 100.
- Distribute photocopiable page 15 for the learners to complete; each pair will also need a set of cards from photocopiable page 9, plus an extra zero.

Plenary

- Ask the learners to describe to a partner what they have been learning about.
- Take feedback. Agree that they have been thinking about multiples of 2, 5, 10 and 100. Establish that they have discovered the rules of divisibility for these numbers.

Success criteria

Ask the learners:

● What number can be divided by 2, 5, 10 and 100? Are there any others? How do you know?

● What is meant by the word multiple?

● What are the rules of divisibility for '2, 5, 10' and 100?

Ideas for differentiation

Support: Provide simple clue cards for these learners so that they can easily identify which numbers can be divided by 2, 5, 10, 100. For example, on one piece of card write '5: 5, 0'.

Extension: During the lesson ask these learners to work out the rules of divisibility for 3 and 9. You may need to suggest they explore the digit totals.

Name: _____

Divided!

You will need:

A partner, a set of 0 to 9 digit cards with an extra zero card.

What to do

● Shuffle the digit cards and decide who will start.

● The first player picks the top three cards and uses them to make a 3-digit number. The aim is to make a number that can be divided by one or more of the numbers 2, 5, 10 and 100.

  You score:

  ● 1 point if it can be divided by 2

  ● 2 points if it can be divided by 5

  ● 3 points if it can be divided by 10

  ● 4 points if it can be divided by 100

  It is possible to score a maximum of 10 points if you can make a number that can be divided by all of 2, 5, 10 and 100.

● Take it in turns to make numbers. Record your results in this table:

Player 1 Name:		Player 2 Name:	
Number	Score	Number	Score
Total score		Total score	

● If you have time, play again. This time make 4-digit numbers. Is it easier to get a higher score with more digits?

Mental calculation strategies 2

- Count on or back in thousands, hundreds, tens and ones to add or subtract. (5Nc8)
- Solve single and multi-step word problems (all four operations); represent them, e.g with diagrams or a number line. (5Pt2)
- Deduce new information from existing information to solve problems. (5Ps4)

Resources

Pencils; paper; set of 0 to 9 digit cards from photocopiable page 9 for each learner; photocopiable page 17.

Starter

- Explain that the learners will be practising adding and subtracting using mental methods.
- Write this calculation on the board: *1456 – 199*. Ask the learners to find the solution. Invite individuals to share strategies. Congratulate any that round and adjust (subtract 200, add one).
- Ask a variety of addition and subtraction calculations that can be easily solved using this strategy, for example: *456 + 199, 5462 – 398*.
- Differentiate according to the attainment levels in your class.

Main activities

- Set this problem: *1546 people went into a store in the morning. 2376 people went in during the afternoon. How many went in altogether?* Ask the learners to talk to a partner about how they would find out.
- Take feedback. Discuss how counting in thousands, hundreds, tens and ones can help. Establish that they could partition one of the numbers and add each part to the other whole number.
- Practise counting forwards and backwards in thousands, hundreds and tens from different starting points, for example 6243, 7029.

- Ask the learners to add and subtract amounts by partitioning the second number, for example:

$$3562 + 1872 \ (+ \ 1000 + 800 + 70 + 2)$$
$$4518 - 2196 \ (- \ 2000 - 100 - 90 - 6)$$

Encourage the learners to make jottings.

- Return to the problem. Keep 2376 whole and then add 1000, 500, 40 and 6, totalling 3922.
- Set similar problems for the learners to solve that involve addition and subtraction.
- Ask the learners to complete photocopiable page 17.

Plenary

- Ask: *Which strategies for addition and subtraction have you been using?*
- Invite the learners to share one of the problems they made up.
- Finish by recapping this strategy. Ask them to add and subtract different three- and four-digit numbers.

Success criteria

Ask the learners:

- How can we add 3982 and 2546? Is there another way? And another?
- How can we subtract 2918 from 4872? Is there another way?
- Can you make up a problem that involves adding or subtracting two numbers? Can you think of another?

Ideas for differentiation

Support: Give these learners three-digit numbers to add and subtract. You could encourage them to draw number lines to help them.

Extension: Give these learners five- or six-digit numbers to add and subtract.

Name: _____

Partitioning

You will need:

A set of 0 to 9 digit cards.

What to do

● Pick four cards and make a four-digit number. Write it down. Rearrange the cards to make another four-digit number.

● Make up a problem that involves either adding or subtracting your two numbers. Solve your problem.

● Do this several times. Make up some addition and some subtraction problems.

● You can record your work here:

Problem	Show how you found the answer

Addition and subtraction 1

Learning objectives

- Use appropriate strategies to add or subtract pairs of two- and three-digit numbers and numbers with one decimal place, using jottings where necessary. (5Nc10)
- Find the total of more than three two- or three-digit numbers using a written method. (5Nc18)
- Solve single and multi-step word problems (all four operations); represent them, e.g. with diagrams or a number line. (5Pt2)

Resources

Pencils; paper; dice; photocopiable page 19.

Starter

- Explain that, to help them during the lesson, the learners will practise adding two- and three-digit numbers using these mental calculation strategies:
  - using doubles and near doubles, for example 36 + 37 (double 36 + 1)
  - bridging through a multiple of 10, for example 84 + 16 (84 + 6 + 10)
  - compensating when adding 9, 19 and so on, for example 245 + 299 (245 + 300 − 1)
  - sequencing, for example 178 + 162 (178 + 100 + 60 + 2)
- Make up different calculations for the learners to answer.
- Ask individuals to explain what they did.

Main activities

- Set this problem: *Assir and Harry were counting the coins they had collected in their money boxes. Assir counted 356 and Harry counted 489. How many did they have altogether?*
- Ask the learners to discuss in pairs how they could answer this problem.
- Take feedback and share ways to add to find the total.
- As a class look at the following possible methods for addition:
  - Sequencing:

356 + 400 + 80 + 9 = 845

- Partitioning:

	3	5	6
+	4	8	9
	7	0	0
	1	3	0
		1	5
	8	4	5

- Compact method, if appropriate:

	3	5	6
+	4	8	9
	8	4	5
	1	1	

- Ask some more problems similar to the one about Assir and Harry.
- Ask the learners to complete photocopiable page 19.

Plenary

- Ask the learners to share with each other the strategies they prefer and why.
- Discuss the other strategies listed above, which might be more appropriate for mental calculation with smaller numbers, for example 134 + 99 (134 + 100 − 1), 125 + 127 (double 125, add 2).

Success criteria

Ask the learners:

- How would you find the total of 356 and 298? Is there another way? And another?
- How could you find the total of 672, 567 and 381? Is there another way?
- How would you describe the sequencing method for addition to someone who didn't know?

Ideas for differentiation

Support: Ask these learners to focus on addition using partitioning. If necessary ask them to work with two-digit numbers.

Extension: Ask these learners to make up four- and five-digit numbers for the photocopiable activity. Encourage them to use their preferred method and then check using an alternative one.

Name: _____

Problems, problems, problems

You will need:

A partner and a dice.

What to do

- Take it in turns to throw the dice six times in total.
  Write each number down as you throw it. Use them
  to make two three-digit numbers.

- Make up an addition problem using these numbers. Solve your problem.

  Here is an example:

  Dice numbers thrown: 5, 3, 3, 6, 2, 4.   Numbers made: 643 and 532.

  Problem: The waiters in a restaurant were unpacking some new plates.
  Freddie unpacked 643. Sal unpacked 532.
  How many plates did they unpack altogether?

  Solution: They unpacked 1175 altogether.

- Record your work here.

Problem 1

Dice numbers thrown: _____   Numbers made: _____

Problem: _____

_____

Solution: _____

Problem 2

Dice numbers thrown: _____   Numbers made: _____

Problem: _____

_____

Solution: _____

- Make up some more problems on the back of this paper.

Addition and subtraction 2

Learning objectives

- Find the total of more than three two- or three-digit numbers using a written method. (5Nc18)
- Check with a different order when adding several numbers or by using the inverse when adding or subtracting a pair of numbers. (5Pt3)
- Solve a larger problem by breaking it down into sub-problems or represent it using diagrams. (5Ps10)

Resources

Pencils; paper; set of cards from photocopiable page 9 for each pair; photocopiable page 21.

Starter

- Explain that, to help them during the lesson, the learners will practise subtracting two- and three-digit numbers using these subtraction strategies:
  - halving, for example $84 \rightarrow 42$ ($\frac{1}{2}$ of $84 = 42$)
  - compensating when subtracting 9, 19 and so on, for example $230 - 198$ ($230 - 200 + 2$)
  - sequencing, for example $345 - 231$ ($345 - 200 - 30 - 1$)
  - complementary addition, for example $286 - 178$ ($178 + 2 + 20 + 86$)
- Make up different calculations for the learners to answer.
- Ask individuals to explain what they did.

Main activities

- Set this problem: *Tyrone and Hussain were counting the number of people at a football match. Tyrone counted 567 in one area and Hussain counted 689 in another. How many did they count altogether? How many more did Hussain count?*
- Ask the learners to discuss in pairs what they would have to do to answer this problem and how they would do this.
- Take feedback and share ways to add to find the total and subtract to find how many more people Hussain counted. Remind the learners of the methods for addition they discussed in the activity on page 18.

- As a class, look at these possible methods for subtraction and relate each to the problem:
  - Complementary addition:

  $567 + 3 + 30 + 89 = 689$, so $3 + 30 + 89 = 122$ more

  - Sequencing:

  $689 - 500 - 60 - 7 = 122$

  - Vertical written method:

	6	8	9
−	5	6	7
	1	2	2

- Discuss how the learners could check their answers: adding in a different order, using another strategy, using the inverse operation.
- Ask some more problems similar to the one about Tyrone and Hussain.
- Ask the learners to complete photocopiable page 21.

Plenary

- Ask the learners to share with each other the strategies they prefer and why.
- Take feedback.
- Discuss the importance of checking and considering whether an answer is reasonable within the context of the problem.

Success criteria

Ask the learners:

- How would you find the difference between 783 and 698? Is there another way? And another?
- How could you check that you are correct? Is there another way to check?
- How would you describe complementary addition to someone who didn't know?

Ideas for differentiation

Support: Ask these learners to focus on subtraction using complementary addition. They may need to work with two-digit numbers.

Extension: Ask these learners to make up four- and five-digit numbers for the photocopiable activity. Encourage them to use their preferred method and then check using an alternative.

Name: _____

Problems, problems, problems 2

You will need:

A partner and a set of 0 to 9 digit cards.

What to do

- Take it in turns to pick six cards in total. Write each number down as you pick it. Use them to make two three-digit numbers.

- Use these numbers to make up a problem involving addition and subtraction. Solve your problem.

  Here is an example:

  Cards picked: 6, 4, 3, 5, 1, 2.   Numbers made: 643 and 512.

  Problem: The cooks in a canteen were making pancakes to raise money for charity. Auzma made 643. Sam made 512. How many pancakes did they make altogether? How many more pancakes did Auzma make than Sam?

  Solution: They made 1155 altogether. Auzma made 131 more than Sam.

- Record your work here.

Problem 1

Cards picked: _____   Numbers made: _____

Problem: _____

_____

Solution: _____

Problem 2

Cards picked: _____   Numbers made: _____

Problem: _____

_____

Solution: _____

- Make up some more problems on the back of this paper.

Cambridge Primary: Ready to Go Lessons for Maths Stage 5 © Hodder & Stoughton Ltd 2012

Multiples and factors

Learning objectives

- Recognise multiples of 6, 7, 8 and 9 up to the 10th multiple. (5Nc5)
- Find factors of two-digit numbers. (5Nc7)
- Use ordered lists and tables to help to solve problems systematically. (5Ps5)

Resources

Pencils; paper; one set of cards from photocopiable page 9 for each pair; photocopiable page 23.

Starter

- Tell the learners they will be practising number sequences to help their reasoning skills which they will need during the lesson.
- Write this sequence on the board: 3, 4, 6, 7. Ask the learners to work out the three numbers that will go after 7 (9, 10, 12). If appropriate, challenge them to work out the three numbers that go before the 3 (–2, 0, 1).
- Encourage them to think of as many possible sequences as they can. Invite volunteers to share their ideas.
- Repeat for other sequences such as 1, 2, 3, 5, 8 (add previous two numbers) and 1, 3, 7, 15 (add 2, 4, 8 ...).

Main activities

- Ask the learners what is meant by the word 'factor'. Agree that it is a number that can be divided into another without a remainder.
- Write 36 on the board. Ask the learners to write all its factors (1, 2, 3, 4, 6, 9, 12, 18, 36). Encourage them to list these in order. Then ask them to pair the factors to make multiplication facts with the answer 36.
- Ask what is meant by the word 'multiple'. Agree that it is a number that is made by multiplying together two other numbers. For example, 35 is a multiple of 5 and 7 because 5 × 7 = 35.
- Ask the learners to write down as many multiples of 5 as they can in a minute, for example 25, 120.

- Remind them of the rules of divisibility they learnt about on page 14: multiples of 2 are even, multiples of 5 end with 5 or 0, multiples of 10 end with zero.
- Ask the learners to work in pairs, taking it in turns to choose a two-digit number and challenge their partner to write down its factors.
- Next, ask the learners to take it in turns to pick a number under 10; their partners write down ten multiples of that number.
- Ask the learners to work through photocopiable page 23.

Plenary

- Ask the learners to describe to a partner what they have been learning about, giving examples.
- Take feedback. Agree that they have been thinking about multiples and factors.
- Call out some numbers and ask the learners to give you their factors or multiples.

Success criteria

Ask the learners:

- What is meant by 'factor'? Can you describe it in another way?
- What are the factors of 64?
- What is meant by the word 'multiple'? Can you describe it in another way?
- What are some of the multiples of 9?
- What operations are involved in finding factors and multiples?

Ideas for differentiation

Support: Ask these learners to focus on multiples of 2, 3, 4, 5 and 10 and factors of numbers within these multiplication tables.

Extension: Ask these learners to focus on multiples of two-digit numbers and factors of three-digit numbers.

Name: _____

Multiples and factors

You will need:

A partner and a set of 0 to 9 digit cards.

What to do

Factors

- One player picks two cards and makes a two-digit number.

- You both then have 30 seconds to write down all the factors of the number. Score one point for each factor you find.

- When you have found the factors of eight numbers, add up your scores.

- The winner is the player with the highest score.

3 6

Multiples

- One player picks a single digit card.

- Both players then have 30 seconds to write down as many multiples of the number as possible. Score one point for each multiple you find.

- When you have found the multiples of eight numbers, add up your scores.

- The winner is the player with the highest score.

5

Multiplication 1

Learning objectives

- Know multiplication and division facts for the 2× to 10× tables. (5Nc3)
- Multiply or divide three-digit numbers by single-digit numbers. (5Nc20)
- Explain methods and justify reasoning orally and in writing; make hypotheses and test them out. (5Ps9)

Resources

Pencils; paper; set of cards from photocopiable page 9 for each learner; photocopiable page 25.

Starter

- Explain that the learners will practise their multiplication tables to help in the lesson.
- Ask the learners to write down all the multiples of 8 to the 10th multiple.
- Invite the learners to call out one of their multiples. Ask the others to give the number which when multiplied by 8 will give this multiple.
- Repeat for other tables.

Main activities

- Set this problem: *Ralph has five sheets of stamps. On each sheet there are 64 stamps. How many stamps does he have altogether?* Ask the learners to discuss in pairs how they might work out the answer.
- Take feedback. Invite the learners to show their methods on the board.
- Aim for these strategies: multiply by 10 and halve, partition, the grid method, the compact method if appropriate. Model any that aren't mentioned by the learners.
- Repeat this for other problems involving two-digit numbers multiplied by a single-digit number.
- Set this problem: *Sonny collects shells. He has six bags each with 346 shells. How many shells does he have altogether?* Tell the learners that to find out the answer they will use the grid method.

- Invite a learner to model the grid method. If no one can do so, then model yourself. Demonstrate drawing the grid. Take each number and explain, for example, that we know 3 × 6 is 18, so 3 × 600 is 1800. When you have all the partitioned answers, find the total.

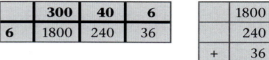

	300	40	6
6	1800	240	36

	1800
	240
+	36
	2076

- Repeat this for similar numbers.
- Ask the learners to complete photocopiable page 25.

Plenary

- Ask the learners to describe to a partner what they have been learning about.
- Take feedback. Agree that they have been thinking about strategies for multiplication.
- Write some calculations on the board for the learners to answer using the grid method.

Success criteria

Ask the learners:

- What strategy could we use to find the answers to a multiplication calculation? Is there another?
- What is the product of 38 and 5? How did you work that out? Is there another way?
- How would you explain how to do the grid method of multiplication to someone who didn't know?

Ideas for differentiation

Support: Ask these learners to focus on multiplying two-digit numbers by a single-digit number using the grid method.

Extension: Ask these learners to focus on multiplying three-digit numbers by a single-digit number.

Name: _____

Multiply!

You will need:

A set of 0 to 9 digit cards.

What to do

● Pick three digit cards and use them to make a number.

● Next pick another card. Multiply the three-digit number you made by this number.

● You can choose which strategy to use.

● Record your work in the table. An example has been done for you.

Numbers picked	Multiplication calculation	Strategy used	Answer
5, 9, 1, 3	591 × 3	<table><tr><td></td><td>500</td><td>90</td><td>1</td></tr><tr><td>3</td><td>1500</td><td>270</td><td>3</td></tr></table>	1773

Multiplication 2

Learning objectives

- Multiply or divide three-digit numbers by single-digit numbers. (5Nc20)
- Multiply two-digit numbers by two-digit numbers. (5Nc21)
- Consider whether an answer is reasonable in the context of a problem. (5Pt7)

Resources

Set of 0 to 9 digit cards from photocopiable page 9 for each learner; pencils; paper; photocopiable page 27.

Starter

- Explain that the learners will practise their multiplication tables to help in the lesson.
- Use digit cards. Select the 6. Hold up the others randomly one at a time. The learners should multiply the number you hold up by 6 as quickly as they can. They write down their answer.
- Repeat this for 7, 8 and 9 times tables' facts.

Main activities

- Set this problem: *Strafford is unpacking boxes of cat food tins in the store. He has 63 boxes. Each box contains 24 tins. How many tins are there altogether?* Ask the learners to discuss in pairs how they might work out the answer.
- Take feedback. Agree that they need to multiply the numbers together. Discuss possible strategies but focus on the grid method.
- If appropriate, invite a learner to demonstrate how to do this, or model the method yourself.
- Ensure that they do it in this way:

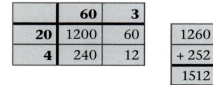

	60	3
20	1200	60
4	240	12

1260
+ 252
1512

- Encourage the learners to explain that, for example, because they know that 6 × 2 is 12, they also know that 60 × 2 is 120 and 60 × 20 is 1200. Discuss whether 1512 is a reasonable answer. Establish that it is using an estimate, for example multiplying 60 by 25 using the mental calculation strategy × 100 and divide by 4.
- Write some multiplications on the board for the learners to solve using this strategy, for example 56 × 32, 87 × 19, 75 × 54. Encourage them to look at their answers and decide if they are reasonable.
- Repeat the Strafford scenario for different sets of numbers or make up your own.
- Ask the learners to complete photocopiable page 27.

Plenary

- Ask the learners to assess how confident they are at using the grid method for multiplication.
- Write some calculations on the board for the learners to answer using this method.
- Discuss the importance of knowing multiplication facts as a tool to help solve problems such as these.

Success criteria

Ask the learners:

- Why are tables facts important?
- What is 46 multiplied by 15? How did you work that out? Is there another way?
- How would you explain how to do the grid method for multiplying two-digit numbers by two-digit numbers to someone who didn't know?

Ideas for differentiation

Support: Ask these learners to focus on multiplying two-digit numbers by a single-digit number using the grid method.

Extension: Challenge these learners to multiply three-digit numbers by two-digit numbers.

Name: _____

Multiplication problems

You will need:

A set of 0 to 9 digit cards.

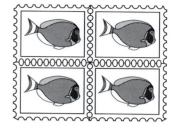

What to do

● Pick two digit cards and use them to make a two-digit number.

● Write the number in the first space in the problems below.

● Make more two-digit numbers and fill in the other spaces.

● Solve the problems.

An example has been done for you.

Problem	Workings	Answer

Shar collected stamps in a stamp album. His album has <u>36</u> pages. On each page there are <u>98</u> stamps. How many stamps does he have altogether?	
	   │ 30 │ 6
	 90 │2700│540
	  8 │ 240 │ 48
		3240 + 288 = 3528

Hattie has _____ bags of shells. Each bag has _____ shells inside it. How many shells does Hattie have altogether?		
Thor is packing boxes of biscuits. In each box there are _____ packets. He has to pack _____ boxes. How many packets does he need?		
The store received _____ crates of cola. In each crate there are _____ cans of cola. How many cans are there altogether?		
The library had a delivery. There were _____ trunks. In each trunk there were _____ books. How many books did the library receive in the delivery?		

● Make up three problems of your own.

● Write them on the back of this paper and solve them.

Cambridge Primary: Ready to Go Lessons for Maths Stage 5 © Hodder & Stoughton Ltd 2012

Division 1

- Multiply and divide any number from 1 to 10 000 by 10 or 100 and understand the effect. (5Nn5)
- Divide three-digit numbers by single-digit numbers, including those with a remainder (answers no greater than 30). (5Nc23)
- Decide whether to round an answer up or down after division, depending on the context. (5Nc26)

Resources

Pencils; paper; dice for each pair; set of cards from photocopiable page 9 for each pair; photocopiable page 29.

Starter

- Explain that the learners will practise multiplying and dividing by 10 and 100.
- Call out some three-digit numbers, for example: *350, 829*. The learners write down what they become when multiplied by 10. They tell you what has happened, for example 'the number is 10 times bigger', 'the digits have moved one place to the left'. Don't accept 'adding a zero'.
- Repeat for dividing by 10 and then for multiplying and dividing by 100.

Main activities

- Set this problem: *Henri picked 145 apples to sell in bags of seven. How many bags will he have?* Ask the learners to discuss in pairs how they can find the answer.
- Establish that they need to divide. Discuss how they might do this; encourage some learners to demonstrate their ideas on the board.
- Focus on grouping. Ask the learners to work in pairs to find the quickest way (take away the greatest number of groups of seven).
- Take feedback. Invite a volunteer to demonstrate if appropriate, or model the method yourself.

145	÷ 7
− 140	(2 lots of 7 is 14, so 20 lots of 7 is 140)
5	

- Say: *Henri can sell 20 bags. What should he do with the rest?* Agree that he should ignore them: they won't fill a bag. Establish that the learners have rounded the answer down.
- Now say: *What should Henri do if he had 145 fish and wanted to put seven in a bowl? How many bowls would he need?* Agree that the answer would need rounding up to 21. The five fish left over would need a bowl.
- Write some divisions on the board for student to solve using grouping, for example 254 ÷ 9, 287 ÷ 8.
- Ask the learners to play the game on photocopiable page 29.

Plenary

- Ask the learners to assess how confident they are at grouping.
- Write some calculations on the board for the learners to answer.
- Discuss the importance of understanding the problem so that they know whether to round up or down.

Success criteria

Ask the learners:

- What is 164 divided by 4? How did you work that out? Is there another way?
- How would you explain grouping to someone who didn't know how to do it?
- How do you know whether to round up or down after a division?

Ideas for differentiation

Support: Ask these learners to focus on dividing two-digit numbers by a single-digit number, taking off multiples of one, five or ten of the divisor.

Extension: Challenge these learners to look for the greatest multiple of the divisor to take away.

Name: _____

Division game

You will need:

0 to 9 digit cards and a dice.

What to do

- Take it in turns to pick three digit cards and use them to make a number. Throw the dice and divide your number by the number thrown on the dice.

- Did you get a remainder? If so, score that number of points.

  For example, if you make 283 and throw 5, the answer will be 56 remainder 3. You score 3 points.

- Record your work in the table below:

Player 1			Player 2		
Cards picked	Calculation	Points scored	Cards picked	Calculation	Points scored
Total score			Total score		

Division 2

Learning objectives

- Recognise odd and even numbers and multiples of 5, 10, 25, 50 and 100 up to 1000. (5Nn13)
- Decide whether to group (using multiplication facts and multiples of the divisor) or to share (halving and quartering) to solve divisions. (5Nc25)
- Choose an appropriate strategy for a calculation and explain how they worked out the answer. (5Ps2)

Resources

Pencils; paper; photocopiable page 31.

Starter

- Explain to the learners that they are going to rehearse odd and even numbers and multiples of 10.
- Call out a two-digit number, for example: *35, 60, 72.* Ask them to clap if it is even, stamp if it is odd, and clap and say 'yes' if it is a multiple of 10.
- Repeat for three- and four-digit numbers.

Main activities

- Set this problem: *Anan had a bag of 84 sweets. She shared the sweets into four equal piles. How many sweets were in each pile?* The learners discuss in pairs what they need to do to solve this.
- Establish that they need to share. Discuss the link between sharing and fractions: the four piles will each have a quarter of the sweets. Discuss how they can find a quarter. Agree that they can halve 84 twice, resulting in 21 sweets per pile.
- Call out amounts and numbers to share them between. Do this within a context that will interest the learners and that is relevant to their lives. For each, the learners write the fraction and its quantity; for example: *24 cakes shared between eight.* The learners write $\frac{1}{8}$ and 3.

- Ask: *What can we do if halving isn't possible?* Establish that they could use a grouping strategy such as sharing equally groups of the divisor.
- Ask more complex questions, for example: *Jasmine put 75 coins equally into three bags. How many coins were in each bag?* The learners should write $\frac{1}{3}$ and 25.
- Ask the learners to complete photocopiable page 31.

Plenary

- Encourage the learners to share some of the problems they made up in the activity.
- Together decide if they are grouping or sharing problems and find the answer.
- Discuss the importance of understanding the problem so that they know whether to group or share.

Success criteria

Ask the learners:

- What is meant by grouping? What is meant by sharing? What is the same about them? What is different?
- Can you make up a problem that involves sharing? Can you make up one that involves grouping?
- What other area of mathematics is sharing linked to?

Ideas for differentiation

Support: Ask these learners to focus on sharing two-digit numbers that they can halve and quarter.

Extension: Challenge these learners to solve problems that involve finding thirds, fifths, sixths and tenths.

Name: _____

Grouping or sharing?

Make up some grouping and sharing problems for your partner to solve. Here is an example of each.

Grouping problem

Wally had 96 DVDs. He stacked them in his DVD cabinet.
Each shelf in the cabinet holds 8 DVDs.
How many shelves did he fill?

Answer: 12 shelves.

Sharing problem

Rashid had 36 cookies. He put them equally in 4 boxes. How many were in each box?

Answer: 9 cookies in each box.

1. Grouping problem _____

    _____

2. Sharing problem _____

    _____

3. Grouping problem _____

    _____

4. Sharing problem _____

    _____

5. Grouping problem _____

    _____

6. Sharing problem _____

    _____

7. Grouping problem _____

    _____

8. Sharing problem _____

    _____

Unit assessment

- What does each of the digits in 45 367 represent?
- How would you round 134 581 to the nearest 10, 100 and 1000?

- Explain how you could add 3425 and 1998.
- Can you make up a problem for 1452 − 1328?
- How would you calculate 145 × 8?

Summative assessment activities

Observe the learners while they take part in these activities. You will quickly be able to identify those who appear to be confident and those who may need additional support.

New numbers

This activity assesses the learner's knowledge of place value.

You will need:

A calculator for each learner (if you don't usually use calculators in your classroom, many mobile phones have a calculator).

What to do

- Organise the learners into groups of four.
- Give each learner a calculator.
- Ask them to key in a number, for example 56.
- For each of the following instructions they add or subtract numbers to or from the previous result. Make sure they don't cancel and type in the new number:
  - Make the 5 become 7 (add 20)
  - Put a 2 in front of the 7 (add 200)
  - Make the 6 into a 3 (subtract 3)
  - Put a 4 in front of the 2 (add 4000)
  - Change the 2 to a 7 (add 500)
- After each instruction ask the learners to show you their calculators so that you can see how they are doing.
- Repeat with different sets of instructions.

Follow-me

This game assesses the learner's ability to recall multiplication facts.

You will need:

A set of cards from photocopiable page 11.

What to do

- Organise the learners into groups of three or four.
- Share out the follow-me cards, keeping one for yourself.
- Read out the multiplication on your card.
- The group with the answer says it and reads out the multiplication on their card.
- Keep going until you read out the final answer.

Distribute copies of photocopiable page 33. Ask the learners to read the questions and write the answers. They should work independently.

Name: _____

Working with numbers

1. Write one hundred and thirty thousand, four hundred and twenty-one in figures.

   _____

2. What is the value of the 6 in 362 879?

   _____

3. Write the missing numbers in this number sequence:

   _____ , 36, 42, _____ , 54, _____

4. Make up a subtraction that will go before the <.

   _____ < 25 + 30

5. Round 345 432 to the nearest:

   10 _____ , 100 _____ , and 1000 _____

6. Divide 356 by 9.

7. Multiply 527 by 5.

8. Add 3567 and 8976.

9. Subtract 1987 from 2438.

10. A pair of jeans costs $58.99 in one shop. The same type costs $65.75 in another. What is the difference in their prices?

    _____

Unit 1B: Geometry and problem solving

Triangles

Starter

- Explain that the learners will rehearse addition and subtraction strategies for two- and three-digit numbers to keep these skills fresh in their minds.

- Ask volunteers to give you ten two- and three-digit numbers to write on the board. Set criteria, for example a three-digit number that ends with 9, a square two-digit number greater than 50.

- Point to pairs of numbers and ask the learners to find their total and difference. Let them make jottings if they wish.

- Take feedback on the different strategies used.

Main activities

- Ask: *What is meant by the word polygon?* Establish that it is any 2D shape with three or more sides. Ask the learners to draw a variety of polygons on plain paper using their rulers.

- Divide the learners into groups of three or four; give the learners a few minutes to describe the shapes they drew to the rest of their group. They should state each shape's name and its properties. Take feedback as a whole class.

- Discuss the fact that the shape name refers to its properties. So, for example, any three-sided shape is a triangle.

- Discuss regular and irregular shapes and perpendicular and parallel sides, drawing shapes to demonstrate.

- Focus on triangles. Invite the learners to sketch any triangles that they have drawn on the board. Ensure that there is a variety, including isosceles, equilateral and scalene.

- Ask the learners to identify similarities and differences. Group the triangles under different criteria, for example perpendicular sides / no perpendicular sides, scalene / not scalene.

- Ask the learners to complete photocopiable page 35.

Plenary

- Invite the learners to describe the triangles they drew in the photocopiable activity.

- Recap the properties of triangles and their classifications. Include perpendicular sides.

Name: _____

Triangles

Use a ruler to draw a triangle to match the clues given and then name it.

Clues	Drawing	Name
All angles acute, all sides different lengths		
All angles the same		
One right angle, two sides the same length		
One obtuse angle, all sides different lengths		
All sides the same length		
One obtuse angle, two sides the same length		

Nets of 3D shapes

- Visualise 3D shapes from 2D drawings and nets, e.g. different nets of an open or closed cube. (5Gs4)
- Deduce new information from existing information to solve problems. (5Ps4)

Pencils; paper; modelling clay; plain paper; rulers; scissors; sticky tape; photocopiable page 37; interlocking squares.

Starter

- Explain to the learners that they will rehearse sharing in relation to fractions to keep these skills fresh in their minds.
- Ask questions such as: *What is half of 32? A quarter of 44? What is three-quarters of 44? What is one-eighth of 24?*
- Ask the learners to write down their answers. Invite volunteers to share their strategies.

Main activities

- Give the learners a piece of modelling clay each. Ask them to make a sphere. Discuss its properties (one curved surface, no edges, no vertices) and where the shape can be seen in real life. Ask them to hold it up in front of them and identify and describe the 2D shape they can see.
- Now ask them to turn their sphere into a cube, describing what they do to a partner (flattening the curved surface). Repeat the questioning used for the sphere. Include the fact that the cube is a prism (same cross-section as the faces at each end).
- Repeat for a cuboid. Expect them to tell you that they are not changing the properties of the cube except for the shape of the faces.
- Repeat for a square-based pyramid.
- Ask the learners to place their pyramids on the table and look down on them from above.

Ask them to visualise what it would look like if the triangular faces were pulled down, and to sketch this on paper.

- Establish they have made a net. The learners cut it out and make a pyramid. Discuss what would make the pyramid more accurate. Agree on measuring to ensure all the triangles are the same size and that the square is a square.
- Ask the learners to make an accurate pyramid.
- Ask the learners to work through photocopiable page 37.

Plenary

- Ask the learners to share how they visualised and made their cubes. Invite volunteers to sketch the nets they created on the board.
- Ask: *What other 3D shapes could you make by visualising and creating nets?* Encourage answers such as cuboids, tetrahedrons, triangular prisms and pentagonal prisms.

Ask the learners:

- What does it mean to visualise? Can you describe this in another way?
- What other shapes can be made using nets? Are there any others?
- How would you recommend other learners attempt to make nets? Which shapes should they try first?

Support: During the photocopiable activity, provide interlocking squares for these learners. If they struggle to visualise, they could use these to make a cube and then open it up.

Extension: During the photocopiable activity, give these learners the opportunity to visualise and then make nets for other 3D shapes, for example tetrahedrons and various prisms.

Name: _____

Cubes

You will need:

Modelling clay, a pencil, paper, scissors and some sticky tape.

What to do

- You have already made the net of a pyramid. Now you are going to make the net of a cube.

- Make a cube out of modelling clay.

- Visualise what it would look like if it was opened up.

- Draw what you think here:

- Now draw your net on a piece of paper, cut it out and make a cube.

- What could you do to make it more accurate?

- Make an accurate net using a ruler to measure the square faces.

- Now cut it out and make an accurate cube.

- There are several possible nets of a cube. Sketch what you think they might be in the space below. You can use your modelling clay cube to help you visualise.

Co-ordinates 1

- Read and plot co-ordinates in the first quadrant. (5Gp1)
- Deduce new information from existing information to solve problems. (5Ps4)
- Explain methods and justify reasoning orally and in writing; make hypotheses and test them out. (5Ps9)

Squared paper; pencils; rulers; photocopiable page 39.

Starter

- Explain that the learners will count in 1000s, 100s, 10s and 1s to keep this skill fresh in their minds.
- Together count in these steps from different numbers, for example: 435, 13 235.
- Write some calculations on the board. Ask the learners to find totals and differences by partitioning the second number and adding each part separately.

Main activities

- Ask: *What is meant by the word co-ordinate? Talk to a partner.* Take feedback. Agree that they are letters, numbers or both that are arranged on a grid to describe the position of things.
- Discuss where these are used in real life, for example maps and atlases.
- Give each learner a piece of squared paper. Ask them to draw a vertical axis on the left of their paper covering ten squares and a horizontal axis from the base of the vertical axis.
- The learners should label both axes from 0 to 10 (on the lines). Model this on the board.

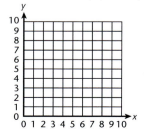

- Ask: *How do we read co-ordinates?* Agree that we read the number on the horizontal axis first and then the vertical.
- Call out various co-ordinates and ask the learners to find them on their grids and mark them with a neat cross. They should list the co-ordinates in the correct format, for example (8, 9).
- Divide the learners into pairs. Ask them to take it in turns to give co-ordinates. They should both plot each point on their grid, then compare their results to see if they agree.
- Ask the learners to work through photocopiable page 39.

Plenary

- Invite volunteers to share what they did in the photocopiable activity. Ask individuals to say the co-ordinates of the crosses from the first grid. Ask others to plot the co-ordinates asked for on a displayed copy of the second grid.
- Use this as an opportunity to check the learners' understanding of reading and plotting co-ordinates.

Ask the learners:

- Can you explain how to plot co-ordinates to someone who doesn't know? Can you explain in a different way?
- When are co-ordinates used in real life? When else?
- What skills are needed to plot and read co-ordinates? Why do you think these are necessary?

Support: Provide a pre-prepared grid for these learners to use during the main activity.

Extension: Ask these learners to draw a 15 × 15 grid and plot crosses onto this.

Name: _____

Co-ordinates

1. Write down the co-ordinates of the crosses.

   a) _____   b) _____   c) _____   d) _____

   e) _____   f) _____   g) _____   h) _____

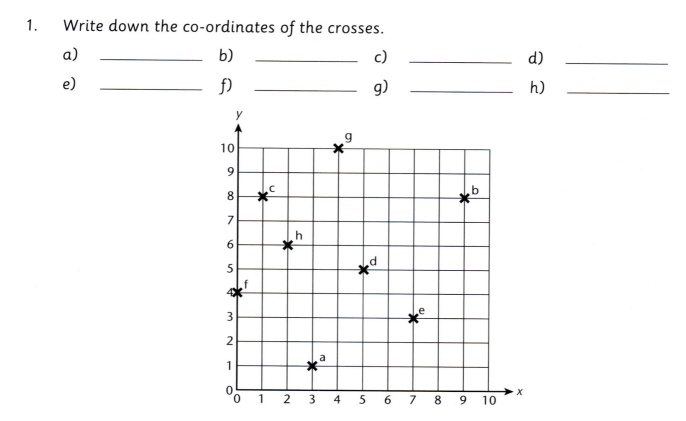

2. Plot the co-ordinates on this grid. Mark each one with a cross.

   (10,10), (6,7), (2,7), (4,6), (7,3), (9,5)

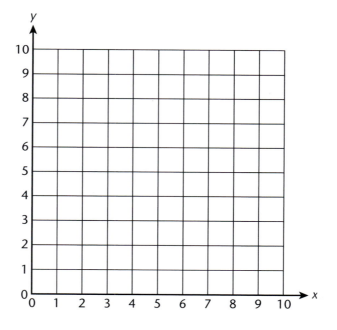

3. Now join the crosses. What shape have you made?

   _____

Co-ordinates 2

Learning objectives

● Read and plot co-ordinates in the first quadrant. (5Gp1)

● Deduce new information from existing information to solve problems. (5Ps4)

● Explain methods and justify reasoning orally and in writing; make hypotheses and test them out. (5Ps9)

Resources

Squared paper; pencils; rulers; photocopiable page 41.

Starter

• Explain that the learners will rehearse multiples to keep these skills fresh in their minds.

• Ask the learners to write down all the multiples of 6 to the 10th multiple.

• Invite the learners to call out one of their multiples. The rest of the class write down the multiplication number sentence to go with it.

• Repeat this for multiples of 7, 8 and 9.

Main activities

• Give each learner a piece of squared paper. Ask them to draw a vertical axis on the left of their paper covering ten squares and a horizontal axis from the base of the vertical axis.

• The learners should label both axes from 0 to 10 (on the lines). Model this on the board.

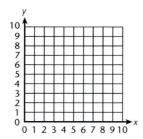

• Ask: *What have you drawn?* Establish that the numbers are the co-ordinates of the grid. Remind them that they read co-ordinates along the horizontal axis first.

• Ask them to plot ten different points on their grid, for example (6, 3), (2, 9). They join these together to form a polygon. What shape have they made? There may be several decagons.

• Ask them to draw another grid. This time give these co-ordinates: *(2, 5), (4, 7), (4, 3)*. Ask them to work out the fourth pair of co-ordinates to make a square and plot the point onto their grid.

• Repeat on the same grid for an isosceles triangle: *(5, 8), (9, 9), (?, ?)* and irregular pentagon: *(6, 1), (5, 2), (7, 4), (9, 2), (?, ?)*.

• Ask the learners to work through photocopiable page 41.

Plenary

• Invite volunteers to share what they did in the photocopiable activity.

• Draw a grid on the board. Plot the co-ordinates for two vertices of an equilateral triangle onto it. Ask the learners to visualise where the third should go. Try out their suggestions.

Success criteria

Ask the learners:

● How do you read co-ordinates? What can you think of to help you to remember this?

● How should you record co-ordinates? Why do you think the brackets are important?

● What helped you to be able to complete the shapes? Is there anything else?

Ideas for differentiation

Support: Provide a pre-prepared grid for these learners to use during the main activity. During the photocopiable activity, ask them to explore different quadrilaterals.

Extension: When these learners have completed the photocopiable activity, ask them to make their own grid to explore irregular quadrilaterals, pentagons, hexagons and octagons.

Co-ordinates 2

You will need:

A partner.

What to do

● Here is a co-ordinate grid:

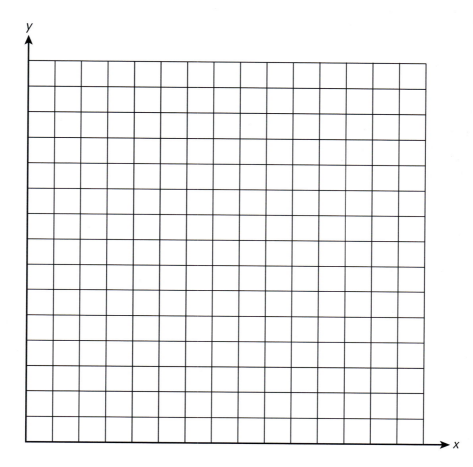

● Label the horizontal and vertical axes from 0 to 15.

1. Plot two sides of a small square on the grid then give your grid to your partner. They need to work out where the other corner should go to make the square. Write down the co-ordinates of all four corners.

2. Plot two corners of a scalene triangle. Your partner then draws the third corner, records the co-ordinates and explains why they placed it where they did.

3. Do this again for a triangle with a right angle and one with an obtuse angle.

Transformations 1

Learning objectives

- Predict where a polygon will be after reflection where the mirror line is parallel to one of the sides, including where the line is oblique. (5Gp2)
- Deduce new information from existing information to solve problems. (5Ps4)

Resources

String; coloured counters; plain paper; mirrors; coloured pencils; photocopiable page 43; squared paper.

Starter

- Explain to the learners that they will rehearse multiplying and dividing any number from 1 to 10 000 by 10 or 100.
- Ask the learners to write down 9, then multiply it by 10 and then 100. *What has happened to the number?* Agree that after multiplication by 10 it is 10 times bigger; the units digit has moved to the tens position and a place holder replaces it in the units position.
- Repeat this for two-, three- and four-digit numbers.
- Call out multiples of 10, 100 and 1000. Ask the learners to divide these by 10 or 100 and to explain what has happened.

Main activities

- Give each pair of learners a piece of string and some coloured counters. Ask them to place the string along the length of their table; this acts as a mirror line. One learner places a selection of counters on one side of the mirror line. The other places identical counters on the other side so that they are a reflection of the first set. Give them a few minutes to record their pattern on plain paper.
- Repeat this, placing the string across the width of their table.

- Repeat again, placing the string diagonally. Discuss the fact that this is more difficult. Ask for suggestions as to what might help them. Establish that they could use a mirror.
- Give each pair a mirror and ask them to use it to check the two sides of their diagonal mirror line are the same.
- Ask the learners to work through photocopiable page 43.

Plenary

- Invite the learners to share what they did in the photocopiable activity. Invite pairs to demonstrate on the board.
- Discuss where they might see reflections in real life, for example, their faces in a mirror, in patterns on wallpaper.

Success criteria

Ask the learners:

- What is a reflection? Is there another way to explain this?
- How could you use a mirror to help you make a reflection? Is there anything else you could do?
- Where do we see reflections in real life? Where else?

Ideas for differentiation

Support: Allow these learners to draw their reflected patterns on squared paper. They should initially become confident in working with horizontal and vertical mirror lines before attempting diagonal ones.

Extension: These learners should focus on diagonal mirror lines.

Reflections

You will need:

A partner, coloured pencils and A4 plain paper.

What to do

- Fold a piece of A4 paper in half.
  You could do this horizontally or vertically.
  The fold will act as a mirror line.

- Decide who will go first.

- The first player draws and colours a small 2D shape on one side of the mirror line.

- The second player matches it on the other side of the mirror line.

- Keep doing this until you have created a pattern.

- Draw your pattern here:

- Do this again on another piece of paper. This time fold the paper diagonally.

- Reflect these shapes across the diagonal mirror lines:

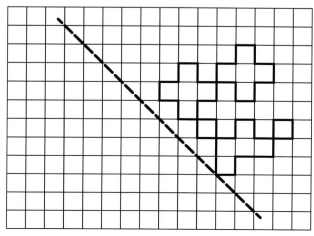

Transformations 2

- Recognise reflective and rotational symmetry in regular polygons. (5Gs2)
- Deduce new information from existing information to solve problems. (5Ps4)
- Explain methods and justify reasoning orally and in writing; make hypotheses and test them out. (5Ps9)

Pencils; paper for recording; plain paper; scissors; coloured pencils; photocopiable page 45; squared paper.

Starter

- Tell the learners that they will rehearse multiplying and dividing any number from 1 to 10 000 by 10 or 100.
- Ask the learners for four-digit numbers. Write them on the board. Ask the learners to multiply them by 10 and write down the new numbers. Now they should divide the original numbers by 10. Decimal numbers are likely to arise. *What is the decimal digit?* (tenth).
- Repeat for five-digit numbers.
- They divide them by 100. *What are the two decimal digits?*

Main activities

- Ask the learners to stand. Call out instructions: *Make a quarter turn to the left, then a three-quarter turn to the right. What is the mathematical term for what you are doing?* Agree that they are rotating in different sized turns. Point out that they are turning from the same point (where they are standing).
- Model the following activity as you tell the learners what to do. First cut a triangular shape out of a piece of paper.
- On another piece of paper, draw a vertical line and a horizontal line through its centre. Position one corner of the triangle where the lines meet and draw around it. Rotate it a quarter of a turn and draw round it again.

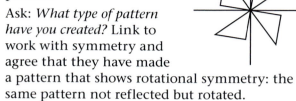

Repeat this twice more. Ask the learners to colour their pattern.

- Ask: *What type of pattern have you created?* Link to work with symmetry and agree that they have made a pattern that shows rotational symmetry: the same pattern not reflected but rotated.
- Discuss the size of the rotations they made. Establish that they are the same size as a right angle and are therefore 90° turns.
- Ask the learners to work through photocopiable page 45.

Plenary

- Invite the learners to share what they did in the photocopiable activity. Pairs should demonstrate both types of rotation on the board.
- Discuss the size of the rotations: a quarter turn is 90°. Half of this is one-eighth or 45°.

Ask the learners:

- What is a rotation? Is there another way to explain this?
- How is a rotation similar to a reflection? How is it different?
- Where do we see rotations in real life?
- If we rotated a shape 180° in a clockwise direction, can you explain where it would be?

Support: Provide these learners with squared paper for rotating; they should focus on rotating shapes by a quarter of a turn.

Extension: Ask these learners to draw their own, more complex shape to rotate.

Rotational symmetry

You will need:

Plain paper, scissors and a pencil.

What to do

● Choose a shape from those at the bottom of the paper and cut it out.

● Draw a cross like this on plain paper.

● Decide which corner you will rotate your shape from. Place that corner at the place where the two lines of your cross meet.

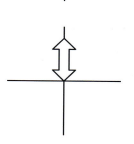

● Draw round it and then rotate it a quarter of a turn, or 90°, and draw round it again.

● Repeat this twice more.

● What happens if you do it again?

● Now choose another shape and cut it out. Draw another cross.

● This time rotate your shape eight times within one full circle.

● How many degrees is each rotation this time?

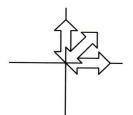

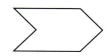

Transformations 3

Learning objectives

● Understand translation as movement along a straight line, identify where polygons will be after a translation and give instructions for translating shapes. (5Gp3)

● Deduce new information from existing information to solve problems. (5Ps4)

Resources

0 to 9 digit cards from photocopiable page 9; pencils; paper for recording; squared paper; counters; photocopiable page 47.

Starter

• Explain that the learners will practise their multiplication tables to keep these skills sharp in their minds.

• Write 8 on the board. Hold up digit cards randomly one at a time. The learners multiply the number you hold up by 8 and write down their answers.

• Repeat for other times tables facts, particularly 6, 7 and 9.

Main activities

• Ask: *What do we mean by a translation? Discuss this with a partner.* Establish that a translation is a movement from one place to another.

• Ask: *When do we see translations?* Examples could include any movement, for example those of people or animals. Ask a volunteer to move from one part of the classroom to another. The class describe this movement using vocabulary such as 'straight', 'right', 'left'.

• Give each learner a piece of squared paper and a counter. Tell them to place the counter in the top-left square. Call out instructions, for example: *move the counter three squares to the right, four squares down, one square to the left.*

• Divide the learners into pairs. One learner in each pair draws a shape in one square and works out a translation to move it somewhere else. They describe the location of their shape and the translation to their partner. Their partner plots the shape and follows the instructions to the new position. They compare their work to see if the instructions were correctly given / followed. They swap roles and repeat.

• Ask the learners to work through photocopiable page 47.

Plenary

• Invite the learners to share what they did in the photocopiable activity. Invite volunteers to describe their translations; the class should follow their instructions to see if they agree.

• Ask the learners to discuss in pairs whether they understand the concept of translation. Take feedback.

Success criteria

Ask the learners:

● Can you make up a definition for the word 'translation'? Is there another possible definition?

● What vocabulary is useful in describing translations? Are there any other words?

● Where do we see translations in real life? Where else?

Ideas for differentiation

Support: During the photocopiable activity, encourage these learners to focus on one shape and find two or three possible ways to translate it.

Extension: During the photocopiable activity, challenge these learners to find two ways to translate each shape and record both.

Translation

On the grid below, five shapes have been translated. Your task is to describe the translations.

1. Draw arrows to show the translations on the grid. You can't go into or across the grey squares!

    An example has been done for you.

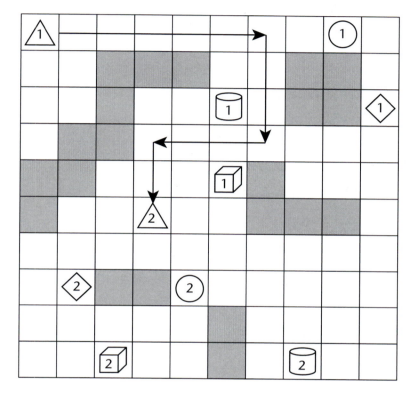

2. Write your instructions below. Keep them simple – for example, use L for left, R for right, D for down.

    My translations:

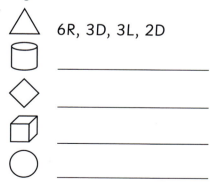

    △   6R, 3D, 3L, 2D

    ▢   _____

    ◇   _____

    ▱   _____

    ○   _____

3. Now add some more shapes or symbols to the grid and translate them.

4. Give the grid to a partner and ask them to write down a possible set of instructions for each of the translations you drew.

Unit assessment

- What is a polygon? Give some examples.
- How would you make a net for a square-based pyramid?
- Explain how you would read co-ordinates on a grid.
- How would you reflect a shape across a mirror line?
- Describe how to rotate a shape a quarter of a turn in a clockwise direction.

Summative assessment activities

Observe the learners while they take part in these activities. You will quickly be able to identify those who appear to be confident and those who may need additional support.

2D shape descriptions

This activity assesses the learners' understanding of polygons.

You will need:

Shape names on card: equilateral triangle, scalene triangle, isosceles triangle, square, rectangle, regular pentagon, irregular pentagon, regular hexagon, irregular hexagon, regular octagon, irregular octagon. Pencils; paper.

What to do

- Organise the learners into groups of four.
- Choose one learner in each group to begin.
- They take a card and describe the shape written on it: they should describe its properties but not give its name.
- The rest of the group draw the shape they think is being described.
- Repeat so that the other learners take a turn at describing a shape.

Which shape?

This game assesses the learners' knowledge of 3D shapes.

You will need:

These 3D shapes in a bag: sphere, cone, cylinder, cube, cuboid, square-based pyramid, triangular prism. Pencils; paper.

What to do

- Organise the learners into groups of three or four.
- Choose a learner to begin.
- They put their hand in the bag and pick a shape.
- Keeping the shape in the bag they describe it according to its properties.
- The other learners write down the name of the shape as soon as they think they know.
- The learner reveals the shape.
- Repeat so that all the learners take a turn at describing.

Distribute copies of photocopiable page 49. Ask the learners to read the questions and write the answers. They should work independently.

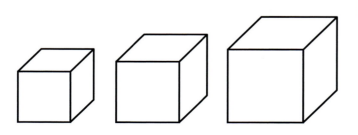

Name: _____

Working with shapes

1. Draw three different types of triangle. Write their names underneath them.

_____   _____   _____

2. Draw a possible net for a square-based pyramid.

3. Plot a fourth point to make a square.

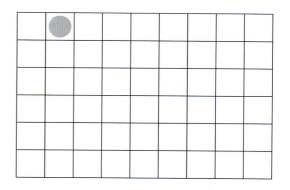

4. Draw where this shape should be after being reflected across the mirror line.

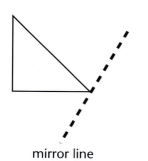

mirror line

5. Translate this shape 6 squares to the right and 5 squares down.

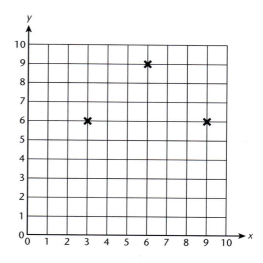

6. Rotate this shape 90° to the left.

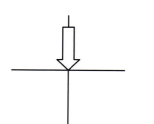

7. Draw where this shape should be after being reflected across the mirror line.

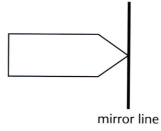

mirror line

Measures

- Convert larger to smaller metric units (decimals to one place), e.g. change 2.6 kg to 2600 g. (5MI2)
- Order measurements in mixed units. (5MI3)
- Round measurements to the nearest whole unit. (5MI4)
- Explain methods and justify reasoning orally and in writing; make hypotheses and test them out. (5Ps9)

Pencils; paper; set of 0 to 9 digit cards from photocopiable page 9 for each learner; photocopiable page 51.

Starter

- Explain that the learners will practise partitioning numbers.
- Write several three-, four- and five-digit numbers on the board. Link these to units of measure, for example 145 cm, 2546 g, 4156 ml, 246 hours. Challenge the learners to partition as many as they can in two minutes.
- Ask individuals to demonstrate what they did on the board.
- Repeat.

Main activities

- Ask: *What is measure? Discuss this with a partner.* Establish that measure includes length, mass, capacity, temperature, time and money.
- Focus on the first three. Write these headings on the board: *length, mass, capacity*.
- Ask: *Can you give me some words that belong under each heading?* Write the words they say under the appropriate heading.
- Focus on units of measure, discussing equivalences. Ask the learners to write these down, for example 10 mm = 1 cm, 100 cm = 1 m, 1000 m = 1 km, 1000 g = 1 kg, 1000 ml = 1 litre.

- Write some mixed units on the board. Ask the learners to write these in another way, for example 235 cm = 2 m 35 cm = 2.35 m, 1456 g = 1 kg 456 g = 1.456 kg. Invite individuals to explain their results. Revise decimal notation; for example, 2 m 35 cm is also 2.35 m because centimetres are hundredths of a metre.
- Write four mixed measures for each of the categories on the board; for example, for length, write: *1456 cm, 1 m 528 cm, 1.4 m, 1 m 48 cm.* Ask the learners to order these from smallest to greatest. Invite volunteers to demonstrate their orders, explaining their thinking.
- Ask the learners to round these measurements to the nearest 10, 100 and 1000.
- Ask the learners to work through photocopiable page 51.

Plenary

- Invite the learners to share the amounts they made in the photocopiable activity. They should show how these can be written in different ways and how they are rounded, as well as the order they placed them in. As a class discuss if these are correct.

Ask the learners:

- What are the different measures we use in everyday life?
- What units do we use to measure length? Mass? Capacity? Are there any others?
- How many millimetres are there in 45 cm? How do you know?
- How else could we write 1 kg 340 g?

Support: During the photocopiable activity, encourage these learners to focus on length. Allow them to use three digit cards to make a three-digit number.

Extension: During the photocopiable activity, ask these learners to use five digit cards to make amounts for length, mass and capacity, not just one.

Measures

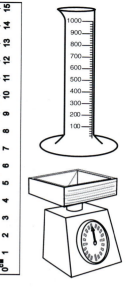

You will need:

A set of 0 to 9 digit cards.

What to do

- Pick four cards and use them to make a four-digit number.

- Decide whether you want to work with length, mass or capacity.

- Write your chosen measure in as many different ways as you can.

  Here is an example:

  > Four-digit number: 4862
  > Measure: 4862 ml, 4 litres 862 ml, 4.862 litres

- Now round it to the nearest 10, 100 and 1000 units.

  Here is an example:

  > 4 litres 860 ml, 4 litres 900 ml, 5 litres

- Change the order of your digit cards to make a new number.
- Write this in as many different ways as you can:

  _____

- Round it to the nearest 10, 100 and 1000 units.

  _____

- Change the order of your digit cards again to make another new number.
- Write this in as many different ways as you can:

  _____

- Round it to the nearest 10, 100 and 1000 units.

  _____

- You have made three measures in total. Order them from smallest to greatest:

  _____

Length

Learning objectives

- Read, choose, use and record standard units to estimate and measure length, mass and capacity to a suitable degree of accuracy. (5MI1)
- Convert larger to smaller metric units (decimals to one place), e.g. change 2.6 kg to 2600 g. (5MI2)
- Draw and measure lines to the nearest centimetre and millimetre. (5MI7)

Resources

A4 paper; pencils; rulers; photocopiable page 53; modelling clay; stopwatches.

Starter

- Explain that the learners will practise counting on and back in steps of different sizes.
- Choose different starting numbers linked to units of length, for example 15 cm. Count on and back from these in steps of 6, 7, 8 and 9.
- Invite volunteers to round a few of the numbers said to the nearest 10 and 100.

Main activities

- Ask: *What can you tell me about length?* Establish that length includes how long, wide or high something is, and the distance from one place to another.
- Ask: *How would we measure length?* Take feedback, for example ruler, tape measure, metre stick, milometer.
- Focus on units of length, discussing equivalences for millimetres and centimetres, centimetres and metres, metres and kilometres.
- Ask questions such as: *How many centimetres are the same as 3 m 45 cm? How do you know?*
- Write four mixed units on the board. Ask the learners to write each in different ways, for example 1 km 540 m = 1540 m = 1.54 km. Invite individuals to explain their thinking.

- Now ask the learners to order these lengths from shortest to longest and then round them to the nearest 10, 100 and 1000.
- Ask them to draw lines of specific lengths that you give on A4 paper, for example 3.7 cm, 12 cm, 9 mm.
- Ask the learners to draw other lines for a partner to measure. They then check to see if their partner is correct.
- Organise the learners into groups of four and ask them to work through photocopiable page 53.

Plenary

- Refer to the photocopiable activity. Take the length of one 'worm' from each group and write it on the board. Ask the learners to order these lengths from shortest to longest.
- Ask the learners to discuss in pairs how confident they are in measuring using centimetres and millimetres.

Success criteria

Ask the learners:

- What instruments can we use to measure lengths?
- What units do we use to measure length? Are there any others?
- How many centimetres are there in 5 m? How do you know?
- How else could we write 12 m 65 cm? Is there another way? Another?

Ideas for differentiation

Support: Encourage these learners to focus on two- or three-digit lengths in metres and centimetres. Allow them to record the length of their 'worms' to the nearest centimetre.

Extension: These learners should work with five- or six-digit numbers for lengths in all four units.

Name: _____

Length

You will need:

Modelling clay and a stop watch.

What to do

- Work in a group of four.

- Make the longest clay 'worm' that you can in 30 seconds.

- Place all four 'worms' in order, from shortest to longest.

- Now estimate and measure each 'worm'. After you have measured the first 'worm', use your knowledge to make a sensible estimate of the second, and so on.

- Record your estimates, measurements and the difference between the two below. Write your measurements in centimetres and in millimetres, for example 12.4 cm, 124 mm.

Name	Estimate	Measurement	Difference

- Did your estimating get better as you worked through the worms?

- Try this again, this time giving yourselves one minute to make your worms!

Name	Estimate	Measurement	Difference

Mass

Learning objectives

- Read, choose, use and record standard units to estimate and measure length, mass and capacity to a suitable degree of accuracy. (5MI1)
- Interpret a reading that lies between two unnumbered divisions on a scale. (5MI5)
- Solve a larger problem by breaking it down into sub-problems or represent it using diagrams. (5Ps10)

Resources

Pencils; paper for recording; weighing scales; photocopiable page 55; modelling clay.

Starter

- Explain that the learners will rehearse place value of five- and six-digit numbers.
- Choose different starting numbers linked to units of mass, for example 42 350 g. Ask: *What does the 4 represent? What about the 5? How else can we write 42 000 g?* Ask the learners to write down their answers.
- Repeat.

Main activities

- Ask: *What can you tell me about mass?* Establish that it is how heavy something is. Share the difference between mass and weight: mass is a measurement of how much matter is in an object, weight is a measurement of how hard gravity is pulling on that object.
- Ask: *How would we measure mass?* Take feedback, for example kitchen and bathroom scales, spring balance.
- Focus on units of mass. Ask questions such as: *How many grams are the same as 3 kg 45 g? How do you know?*
- Write four mixed units on the board. Ask the learners to write each in different ways, for example 5 kg 250 g = 5250 g = 5.25 kg. Invite individuals to explain their thinking.

- Now ask the learners to order these measurements from lightest to heaviest and then round them to the nearest 10, 100 and 1000.
- Divide the class into groups of four, and ask each group to collect five items from around the classroom. They feel them and order from lightest to heaviest.
- Tell the groups to estimate and then weigh their lightest item using a set of scales. They should compare their estimates with the actual mass, then use this knowledge to estimate the second item. They continue until they have estimated and weighed all the items.
- Ask the learners to complete photocopiable page 55 in their groups.

Plenary

- Refer to the photocopiable activity. Invite groups to share what they did. They should explain when and why they made estimates. Check to see if everyone agrees on the total mass of all the fruit (2760 g = 2 kg 760 g = 2.76 kg).

Success criteria

Ask the learners:

- How many grams are there in 4 kg? How do you know?
- How else could we write 15 kg 250 g? Is there another way? Another?
- What is an estimate? Can you explain this in a different way?
- How can you make a sensible estimate?

Ideas for differentiation

Support: Ensure groups of mixed attainment so that these learners can be supported by their peers.

Extension: Challenge these learners to work with six- and seven-digit masses.

Name: _____

Mass

You will need:

Modelling clay, weighing scales, paper and a pencil.

What to do

● Work in a group of four.

● Jamal's uncle teaches grade 1 at another school. He would like you to use modelling clay to make some shapes that look like fruit for his class. He wants them to weigh 150 g, 270 g, 420 g and 540 g.

● You need to make two of each. When you have done this, make a label for each fruit to show its mass.

● Reflect on what you did. When did you estimate? How did you get the correct mass?

● The scales below weigh up to 1 kg. Show the weight of the fruit on the scales.

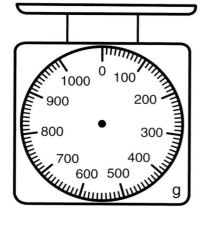
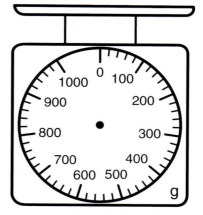

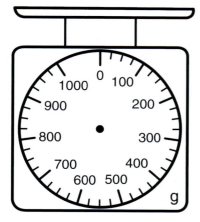
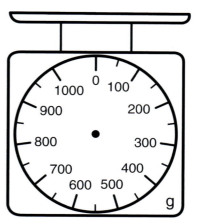

● What is the total weight of all eight fruit? Write this in three different ways.

_____

_____

Capacity

- Read, choose, use and record standard units to estimate and measure length, mass and capacity to a suitable degree of accuracy. (5MI1)
- Convert larger to smaller metric units (decimals to one place), e.g. change 2.6 kg to 2600 g. (5MI2)
- Compare readings on different scales. (5MI6)

Pencils; paper; different containers, for example 500 ml, 1 litre, 2 litre bottles; selection of different measuring jugs and cylinders; water; photocopiable page 57; set of cards from photocopiable page 9 for each pair.

Starter

- Explain that the learners will order amounts from smallest to largest.
- Write a selection of mixed units of length on the board. Ask the learners to write them in order. Ask: *How do you know you have the correct order?*
- Repeat for mass and capacity.

Main activities

- Ask: *What can you tell me about capacity?* Establish that it is the amount a container holds. Share the difference between capacity and volume: capacity is the amount a container can hold, while volume is the amount of a substance inside the container.
- Ask: *How would we measure capacity?* Take feedback, for example measuring cylinders of different shapes and sizes.
- Focus on units of capacity. Ask questions such as: *How many millilitres are the same as 2.5 litres? How do you know? When might we measure in millilitres? What about litres?*
- Write four mixed units on the board. Ask the learners to write each in different ways, for example 3 litres 125 ml = 3125 ml = 3.125 litres. Invite individuals to explain their thinking.

- Now ask the learners to order these measurements from least to most and then round them to the nearest 10, 100 and 1000.
- Show a selection of different containers. Invite the learners to order the containers from least to most in terms of capacity. Ask: *How could we find out exactly how much these hold?* Establish that they could fill them with water and use a measuring jug to find out how much is in each. Invite volunteers to do this.
- Order the different capacities on the board.
- Ask the learners to work through photocopiable page 57.

Plenary

- Refer to the photocopiable activity. Invite groups to explain what they did. Ask them to write the amounts they made on the board and share their total. The class checks to see if they were correct.
- Assess how they feel about their understanding of capacity.

Ask the learners:

- How many millilitres are there in 3 litres? How do you know?
- How else could we write 16 litres 120 ml? Is there another way? Another?
- When do people use or see litres and millilitres in real life? Can you give me other examples?

Support: Ensure these learners are paired with a higher-attaining learner so they can be supported.

Extension: Challenge these learners to work with six- and seven-digit capacities.

Capacity

You will need:

A partner, 0 to 9 digit cards, a measuring cylinder, two or three containers and water.

What to do

- Pick three digit cards and make a three-digit number. Pretend that this is the capacity of a container in millilitres.

- Write your capacity here: _____

- Now make three more capacities and write them next to the first.

- Pour water into different containers, estimating to match your four capacities.

- Now measure the water you poured using a measuring cylinder. How close were your estimates?

- Show the four capacities on these measuring cylinders:

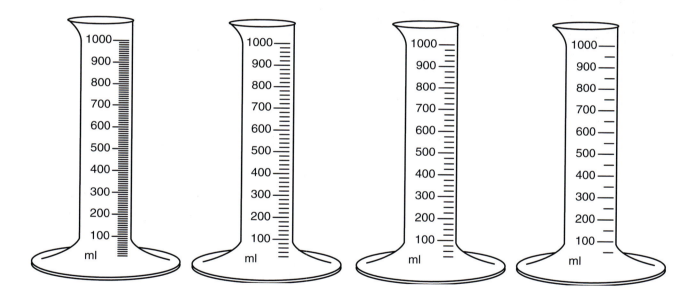

- What is the total capacity of all four containers? Write this in three ways.

_____

_____

_____

Time 1

Learning objectives

● Recognise and use the units for time (seconds, minutes, hours, days, months and years). (5Mt1)

● Tell and compare the time using digital and analogue clocks using the 24-hour clock. (5Mt2)

● Understand everyday systems of measurement in ... time and use these to perform simple calculations. (5Pt1)

Resources

Pencils; paper; class clock; small clock faces; photocopiable page 59 for each group; scissors.

Starter

• Explain that the learners will be thinking about the units for time, to help them in the lesson.

• Ask questions such as: *How many minutes in an hour? Two hours? Half an hour? Four and a half hours?* Ask the learners to write down their answers. Invite individuals to share their strategies.

• Repeat for seconds in different numbers of minutes.

Main activities

• Ask: *What can you tell me about time? Talk to your partner and think of five things to share.* Take feedback. Discuss the units used for telling the time, for example seconds, minutes, days, months, years. Establish that we use digital and analogue clocks. Digital time can be told in 12 hours; a.m. or p.m. annotations are needed to show what part of the day it is. It can also be told in 24 hours.

• Ask the learners to tell you the different places that they have seen 24-hour time, for example mobile phone, computers, ovens.

• Give each pair of learners a clock face. Ask them to show you different times. They should write down the equivalent digital time. For example, say: *20 minutes past two in the afternoon.* The learners show that time on their clock face and write down 2:20p.m. and, if appropriate, 14:20.

• Set problems such as: *The school bus arrived at 8:10a.m. The journey to school took 45 minutes. Show me on your clocks the time it got to school.*

• Ask: *How can we convert 12-hour clock time to 24-hour?* Take feedback and then demonstrate using a number line:

```
   12:00 1:00 2:00 3:00 4:00 5:00 6:00 7:00 8:00 9:00 10:00 11:00
◄──────────────────────────────────────────────────────────────
   00:00 01:00 02:00 03:00 04:00 05:00 06:00 07:00 08:00 09:00 10:00 11:00

   12:00 1:00 2:00 3:00 4:00 5:00 6:00 7:00 8:00 9:00 10:00 11:00
──────────────────────────────────────────────────────────────►
   12:00 13:00 14:00 15:00 16:00 17:00 18:00 19:00 20:00 21:00 22:00 23:00
```

• Use this to practise adding and subtracting hours and minutes.

• Divide the class into groups of two or three and ask them to play the game on photocopiable page 59.

Plenary

• Ask the learners to give you a variety of analogue times. The class write down the 12- and 24-hour digital equivalences.

• Ask them to assess how confident they are at telling these types of time.

Success criteria

Ask the learners:

● What is another way to say 14:55? Can you tell me in another way?

● Where would the hands on an analogue clock point if the time is 8:35a.m.?

● What is the same about 12- and 24-hour clock time? What is different?

Ideas for differentiation

Support: Ask these learners to focus on five minute intervals of time in both analogue and digital time.

Extension: After playing the game once, these learners can make their own set of cards with times to the minute.

Match it!

This game is for two or three players.

1. Cut out the cards. Shuffle them well. Deal out four to each player. Place the rest in a pile face down on the table.

2. The aim is to collect three cards that show the same time. Take it in turns to ask any player for a time to match one that you already have. If they have it, they give it to you. If they don't, pick a card from the pile.

3. The winner is the player with the most trios when all the cards have been used.

2:55	2:35	9:50	8:05
6:10	3:35	10:50	2:55
Five minutes to three	Twenty-five minutes to three	Ten minutes to ten	5 minutes past 8
Ten minutes past six	Twenty-five minutes to four	Ten minutes to eleven	Five minutes to three

Time 2

Learning objectives

- Read timetables using the 24-hour clock. (5Mt3)
- Use a calendar to calculate time intervals in days and weeks (using knowledge of days in calendar months). (5Mt5)
- Choose an appropriate strategy for a calculation and explain how they worked out the answer. (5Ps2)
- Use ordered lists and tables to help solve problems systematically. (5Ps5)

Resources

Photocopiable page 192; pencils; paper; photocopiable page 61; small clock faces.

Starter

- Explain that the learners will be thinking about the units for time to help them in the lesson.
- Give each pair of learners photocopiable page 192. Ask questions such as: *How many days in June? Saturdays in August? How many weeks in September and October? How many weeks and days from 2nd March to 5th April?* Ask the learners to write down their answers. Invite individuals to explain how they found some of the answers.

Main activities

- Draw this time number line on the board:

12:00 1:00 2:00 3:00 4:00 5:00 6:00 7:00 8:00 9:00 10:00 11:00
00:00 01:00 02:00 03:00 04:00 05:00 06:00 07:00 08:00 09:00 10:00 11:00

12:00 1:00 2:00 3:00 4:00 5:00 6:00 7:00 8:00 9:00 10:00 11:00
12:00 13:00 14:00 15:00 16:00 17:00 18:00 19:00 20:00 21:00 22:00 23:00

- Use it to practise adding and subtracting hours and minutes, for example $06:00 + 4\frac{1}{2}$ hours.
- Ask: *What is a timetable?* Take feedback. Establish that a timetable gives information about, for example, when buses arrive and depart. Ask the learners where they have seen timetables, for example TV guides, train / plane arrivals and departures.

- Give each learner a copy of photocopiable page 61. Focus on bus 1. Ask the learners to tell you what time it leaves various towns. Ask questions related to the time it takes to get from one place to another, for example from the shopping mall to the mosque, from the hospital to the park. The learners write down their answers.
- Repeat for bus 2.
- Ask: *What do the spaces in the timetable mean?* Agree that it means that the buses don't stop at these places.
- Ask the learners to work out the time it takes each bus to go from bus stop to bus stop. They should also find the total time and work out the shortest journey time and the longest.

Plenary

- Refer to the timetable again. Take feedback from the activity. Ask: *Which bus had the shortest total journey time? Which had the longest?*
- Ask the learners to make up some questions about the timetable to ask the rest of the class.

Success criteria

Ask the learners:

- What is the time difference between 10:45 and 13:15? How did you work that out? Is there another way?
- If bus 3 was running 25 minutes late, what time would it get to the zoo? How do you know?
- If I arrived five minutes early to catch bus 6 at the school, what time did I get there?

Ideas for differentiation

Support: Ask these learners to focus on buses 3 and 4. Draw a time number line for the appropriate times to help them. If they prefer they could use clock faces.

Extension: Ask these learners to focus on buses 5 and 6.

Name: _____

Timetable

This timetable shows the times that different buses leave the bus station and arrive at the zoo.

	Bus 1	Bus 2	Bus 3	Bus 4	Bus 5	Bus 6
Bus station	08:00	08:45	09:05	09:15	10:21	10:47
Shopping mall	08:20	09:00	09:35	09:30	10:47	10:52
Hospital	08:45	09:30		10:00		11:38
School	09:05	10:00	10:20	10:30	12:06	12:19
Mosque	09:55		10:55	11:00	12:42	12:53
Park	10:15		11:35	11:15		13:11
Zoo	10:45	11:45	12:20	12:00	13:36	13:52

Make up some questions about the timetable to ask the rest of the class.

_____

_____

_____

_____

_____

_____

_____

Perimeter and area 1

● Measure and calculate the perimeter of regular and irregular polygons. (5Ma1)
● Understand area measured in square centimetres (cm²). (5Ma2)
● Deduce new information from existing information to solve problems. (5Ps4)
● Consider whether an answer is reasonable in the context of a problem. (5Pt7)

Resources

Pencils; paper for recording; centimetre squared paper; photocopiable page 63.

Starter

• Explain that the learners will be rehearsing mental addition and subtraction strategies.
• Write two digits on the board. Ask the learners to use them to make two two-digit numbers, then find their total and their difference.
• Repeat with other two- and then three-digit numbers.

Main activities

• Ask: *What do we mean by perimeter and area? Talk to your partner.* Take feedback. Agree that area is the amount of space something takes up, perimeter is the outside of that area.
• Ask the learners to give examples, for example a garden surrounded by a fence. The area is the amount of garden, the fence is the perimeter.
• Ask: *How do we measure perimeter and area?* Establish that the perimeter of something is a length and so will be measured in millimetres, centimetres, metres and kilometres. Area is measured in square millimetres, centimetres, metres and kilometres.
• Ask: *What perimeters and areas might be measured in each of these units?* The learners should share ideas with a partner first and then with the class.

• Give each learner a piece of centimetre squared paper. Ask them to draw a square of any size. Discuss how they can find the perimeter and area. For the perimeter: find the length of the sides and multiply by 4. For the area: count the squares.
• Ask the learners to find the perimeters and areas of their squares and give them to a partner to check.
• Ask them to explore perimeters and areas of other squares and rectangles.
• Ask the learners to work through photocopiable page 63.

Plenary

• Discuss the problem on photocopiable page 63. Ask the learners for their thoughts. Invite volunteers to share examples of when the area gets bigger and perimeter doesn't and vice versa. Agree that the statement made by Sanjit is **sometimes** true.
• Ask the learners to assess their confidence in this area of mathematics.

Success criteria

Ask the learners:

● What can you tell me about perimeter? What else?
● What can you tell me about area? What else?
● If I draw a square which has sides of 4 cm, what is its perimeter? How did you work that out? What is its area? How do you know?

Ideas for differentiation

Support: Allow these learners to add side lengths to find perimeters and count squares to find areas.

Extension: These learners should look for quick ways to find perimeters, for example 2 × (*l* + *w*) and areas, for example length × width.

Name: _____

Perimeter and area

Sanjit thinks...

> If the area of something gets bigger so does the perimeter.

1. What do you think? Is what Sanjit thinks:

    ● always true

    ● sometimes true

    ● never true?

2. Use the space below to explain your thinking.

3. Draw squares and rectangles on squared paper as part of your explanation. You can stick them here.

Perimeter and area 2

- Measure and calculate the perimeter of regular and irregular polygons. (5Ma1)
- Understand area measured in square centimetres (cm²). (5Ma2)
- Use the formula for the area of a rectangle to calculate the rectangle's area. (5Ma3)
- Deduce new information from existing information to solve problems. (5Ps4)

Pencils; paper for recording; centimetre squared paper; plain paper; photocopiable page 65.

Starter

- Explain that the learners will be rehearsing mental addition and subtraction strategies.
- Call out two two-digit numbers. Ask the learners to find their total and their difference, making jottings if they need to. Invite volunteers to share their strategies.
- Repeat with other two- and then three-digit numbers.

Main activities

- Recap area and perimeter: area is the amount of space something covers, perimeter is the distance round the outside of that area. Perimeter is measured in different units of length, area is measured in squares of those units.
- Set this problem: *Hammed wants to cover his back yard with grass. His back yard measures 12 m by 10 m. What area will he cover?*
- Give each learner a piece of centimetre squared paper and ask them to draw a rectangle to represent the scaled-down size of the back yard (12 cm by 10 cm). Discuss how they can find the area. Agree that they can count the squares. They could also multiply 12 cm by 10 cm.
- Ask: *What is the perimeter of Hammed's back yard?* Discuss ways of finding this, for example: add the length and width, then double, that is $2(l + w)$; or double the length, double the width and add, that is $2l + 2w$.

- Sketch an equilateral triangle and a regular pentagon. Ask: *How can we find the perimeter of these shapes?* Agree that they could measure the length of one side and multiply it by 3 or 5. Ask: *What would we do for irregular shapes?* Establish that they would measure all the sides and find the total.
- Ask the learners to draw irregular polygons on a piece of plain paper, and find their perimeters.
- Ask the learners to work through photocopiable page 65.

Plenary

- Discuss the work from the photocopiable activity. Ask: *What do you notice about the perimeters of the different tetrominoes?* Agree that the area is the same but the perimeters are different.
- Establish that area and perimeter don't have a relationship with each other.

Ask the learners:

- Is this statement correct: if the area gets bigger so does the perimeter? Why not?
- Can you give an example of when the area stays the same and the perimeter is different?
- If I draw a regular hexagon which has sides of 8 cm, what is its perimeter? How did you work that out?

Support: Allow these learners to draw their shapes on squared paper. They can add side lengths to find perimeters and count squares to find areas.

Extension: When they have completed photocopiable page 65, challenge these learners to explore the different shapes they can make out of five squares (pentominoes).

Polygons

1. Work out the perimeters and areas of these shapes. They are all made from squares of side 1 cm.

    This is a domino:

    Area: _____

    Perimeter: _____

    These are triominoes:

    Area: _____   Area: _____

    Perimeter: _____   Perimeter: _____

2. Tetrominoes are made from four squares. How many can you make using squares of side 1 cm? There are five altogether. Draw them here:

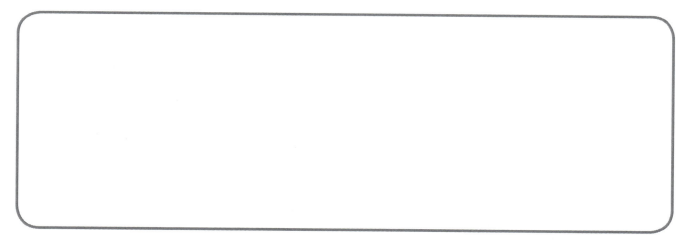

3. For each, work out the perimeter and the area. Write each beside the correct shape.

4. What do you notice about their areas?

    _____

5. What do you notice about their perimeters?

    _____

Unit assessment

- Describe the different things that we measure.
- What units can we use to measure length?
- What equipment can we use to measure length?
- What is meant by perimeter?
- How can we measure area?

Summative assessment activities

Observe the learners while they take part in these activities. You will quickly be able to identify those who appear to be confident and those who may need additional support.

How heavy?

This activity assesses the learners' understanding of mass.

You will need:

Five objects from around the classroom for each group; 1 kg bag of rice or similar; pencils; paper; weighing scales.

What to do

- Organise the learners into groups of four.
- Give each group five objects.
- They feel each one and put them in order from lightest to heaviest.
- They then compare each object with the bag of rice and estimate how heavy they think each object is.
- They write a label for each weight.
- They weigh the objects on the scales to check their estimates.

What capacity?

This activity assesses the learners' knowledge of capacity.

You will need:

Three containers for each group; 2 litre pop / water bottle; pencils; paper; water; measuring cylinders.

What to do

- Organise the learners into groups of four.
- Put the containers and the 2 litre pop / water bottle in front of them.
- The learners look carefully at each container, compare it with the bottle and estimate the capacities of the containers. They write these down.
- Invite the learners to fill the containers with water to find out the actual capacities.
- They compare the actual capacities with their estimates to see how close they were.

Distribute copies of photocopiable page 67. Ask the learners to read the questions and write the answers. They should work independently.

Name: _____

Working with measures

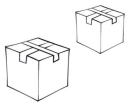

1. Write 2.45 litres in two different ways.

   _____

2. Write 25 cm 5 mm in two different ways.

   _____

3. What measuring instruments could we use to measure mass?

   _____

4. A parcel weighs 5 kg 250 g. Another weighs 5.2 kg.

   Which is the heaviest? How do you know?

   _____

5. Write down ten units that we use to measure time.

   _____

   _____

6. On the back of this paper, draw a clock face to show 20 minutes past 8 in the morning. Write the time beside it digitally.

7. Sharma went for a walk. She left home at 3:15 pm and got back at 4:35 pm. How long was she out walking?

   _____

8. Use a time number line to work out the difference between 12:20 and 14:40.

   _____

9. I draw a square with sides of 12 cm.

   a) What is its perimeter? _____

   b) What is its area? _____

10. Draw a shape with a perimeter of 24 cm on the back of this paper.

    What is its area?

Unit 2A: Number and problem solving

Numbers to 1 000 000

● Count on and back in steps of constant size, extending beyond zero. (5Nn1)

● Partition any number up to one million into thousands, hundreds, tens and units. (5Nn3)

Resources

Counting stick; pencils; paper; set of 0 to 9 digit cards from photocopiable page 9 for each learner; photocopiable page 69.

Starter

• Explain that as numbers are the focus of the lesson, the learners will practise counting in steps of different sizes.

• Show the counting stick. Explain: *One end is zero, the other is 90.* Ask: *What steps should we count in to get from zero to 90?* Agree on nine. Together, count from zero to 90 and back. Place your finger on different divisions. Ask: *What number would be found here? Here? How many lots of nine is this?* Put your finger on the second interval. This is zero. Together, count in steps of nine from zero to one end (81). Then count back to the other (−9).

• Repeat, positioning zero in different places.

Main activities

• Write 3427 on the board. Ask the learners to partition it and write a number sentence to show this: 3000 + 400 + 20 + 7.

• Repeat with other numbers.

• Give the learners a set of digit cards each. Ask them to make 34, then 134, 1324, 13 624 and finally 713 624. Say all these numbers using words.

• Focus on 713 624. Ask: *Which digit shows us how many hundreds / tens of thousands / units there are? Swap the 7 and 1. Is the number bigger or smaller? Roughly how much bigger? Swap the 1 and 6. Roughly how much bigger is it now?* Invite someone to read the new number.

• Repeat this, swapping other digits. Then repeat from the beginning, making a different number.

• Give each learner a copy of photocopiable page 69 and ask them to play the game.

Plenary

• Set this problem: *I have a number: it has a 1 in the tens of thousands position, 4 in the tens, 6 in the ones of thousand, 7 in the ones and 9 in the hundreds. What is my number?*

• Repeat with other similar problems.

Success criteria

Ask the learners:

● What three-digit number can be made with the digits 4, 9 and 7? How many could you make?

● What number could you make with the digits 4, 6, 2, 9 and 1? Is there another?

● What can you tell me about the number 386 102? What else?

● What do we call the zero? What is its job?

Ideas for differentiation

Support: Prepare a grid with the headings 100, 10, 1 for these learners to use during the photocopiable activity and ask them to make three-digit numbers.

Extension: During the photocopiable activity, ask these learners to work out how much greater the largest number made is. They add an extra column to the table to record this.

Name: _____

Big numbers!

You will need:

A partner and a set of 0 to 9 digit cards each.

What to do

- You are going to use the digit cards to make six-digit numbers. The aim is to make the largest number you can.

- Shuffle the cards and place them face down on the table in front of you.

- Take it in turns to pick a digit.

- Decide which position you want your digit to go in and write it in the table. Keep doing this until you have made a six-digit number. Write the full number in the last column.

- When you have both made a six-digit number, compare your number with your partner's. The player with the highest number scores a point.

- The player with the highest score at the end is the winner.

	100 000	10 000	1000	100	10	1	Number
1							
2							
3							
4							
5							
6							
7							
8							

- Draw another grid on the back of this paper. Play again – but this time aim to make the smallest six-digit numbers you can.

Number sequences

Learning objectives

- Know by heart pairs of one-place decimals with a total of 1, e.g. 0.8 + 0.2. (5Nc1)
- Derive quickly pairs of decimals with a total of 10, and with a total of 1. (5Nc2)
- Recognise and extend number sequences. (5Nn12)
- Explore and solve number problems and puzzles, e.g. logic problems. (5Ps3)

Resources

Pendulum (three interlocking cubes on a piece of string or similar); pencils; paper; photocopiable page 71; set of digit cards from photocopiable page 9 for each learner.

Starter

- Tell the learners they will be finding decimal number pairs that total 1 and 10.
- Use the pendulum. As it swings one way call out a decimal number, for example 0.7 or 0.1. As it swings the other way, the learners should call out the number that goes with it to total 1 (0.3, 0.9).
- Repeat for pairs of decimal numbers that total 10, for example 3.5 (6.5), 8.4 (1.6).

Main activities

- Ask the learners to give you three single-digit numbers. Write them in order on the board. Ask them to work out a sequence from these numbers. It can use any rule, for example 2, 5, 9: + 3, + 4 or double + 1, double − 1. They then continue or repeat the rule, so extending the sequence.
- Invite individuals to explain their sequences.
- Repeat with other numbers.
- Write this sequence on the board: 25, 50, 75, 100. Ask the learners to continue it on to 125 and back to −125.

- Ask the learners to discuss this question in pairs: *What is the tenth number in the sequence going up from 25?* Take feedback and establish they would multiply 25 by 10. Repeat for other numbers.
- Discuss how they could find any positive number in the sequence: multiply the chosen number by 25. Challenge the learners to find numbers of their choice within this sequence, working in pairs.
- Repeat for other sequences.
- Ask the learners to work through photocopiable page 71.

Plenary

- Take feedback from the photocopiable activity.
- Invite pairs to share the sequences they made. Ask the class to work out the rule and to extend the sequence further. The pairs say whether the class are right and explain if not.

Success criteria

Ask the learners:

- What is a sequence? Can you explain in a different way?
- Can you give me a sequence? What is your rule?
- My sequence is 2, 5, 11, 23. What is the rule I have made up? What are the next three numbers? How do you know? Is there another possible rule?

Ideas for differentiation

Support: During the photocopiable activity ask these learners to make a two-digit number and use the third to devise their rule, for example adding that number repeatedly.

Extension: During the photocopiable activity ask these learners to make a two-digit starting number, then use another two-digit number to devise their rule.

Name: _____

Sequences

25, 50, 75, …?

4, 6, 10, …?

You will need:

A set of 0 to 9 digit cards.

What to do

- Pick three cards and lay them on the table in front of you.

- Use the numbers on the cards to make a sequence.

- You can choose how you use the numbers:

  a) You could use them as single digits.

  For example: I pick 3, 6, 1.

  I make this sequence: 1, 3, 6, 10, 15 and so on.

  Each time I add on a number that is one bigger than the previous number I added on.

  b) You could make a two-digit number and use the third to create your sequence.

  For example: I pick 7, 2, 9.

  I make 72 and then make a sequence by adding 9 each time.

  c) You could make a three-digit starting number.

  For example: I pick 8, 4, 6.

  I use 846 as my starting number and add 150 each time.

  d) You could do something else – it's up to you!

- Write your numbers below:

  _____

- Now make up your sequence:

  _____

- What was your rule?

  _____

- Could there be another rule?

  _____

10, 100 and 1000

- Know multiplication and division facts for the 2× to 10× tables. (5Nc3)
- Multiply and divide any number from 1 to 10 000 by 10 or 100 and understand the effect. (5Nn5)

Pencils; paper; photocopiable page 73; set of 0 to 9 digit cards from photocopiable page 9 for each learner.

Starter

- Explain that the learners will practise their multiplication and division facts.
- Call out multiples of 6, 7, 8, 9 and 10, for example 56, 100. Tell the learners to write down one multiplication and one division number sentence to go with them, for example 7 × 8 = 56, 56 ÷ 8 = 7.
- Ask the learners to show their answers so that you can check them.

Main activities

- Write this on the board: × 10, × 100, × 1000. Ask the learners to discuss in pairs what happens to a number if these operations are applied. Take feedback. Agree that the number will be 10, 100 and 1000 times bigger.
- Write 72 on the board. Invite volunteers to tell you what happens to the position of the digits if it is multiplied by 10 (move one place to the left). Ensure that you talk about zero as the place holder holding the place of the units. Repeat for 100 and 1000.
- Call out a variety of two-, three- and four-digit numbers. The learners should multiply each by 10, 100 and 1000 and write down the results. Invite individuals to read out their answers.

- Repeat for division. Write 130 on the board. Ask: *What happens if I divide this by 10?* Agree that the result is ten times smaller; each digit moves a place to the right. Repeat for 125. Discuss what happens to the 5 (it becomes 5 tenths).
- Ask the learners to work through photocopiable page 73.

Plenary

- Ask: *What have you been learning about in today's lesson?* Take feedback.
- Invite individuals to share the numbers they made, read what they became when multiplying and dividing, and explain what happened to the numbers each time.

Ask the learners:

- Can you explain what happens when you multiply a number by 10? What about 100? What about 1000?
- What happens when you divide a number by 10? What about 100?
- If there are digits after the decimal point, what do these represent?

Support: During the photocopiable activity ask these learners to make two-digit numbers and multiply them by 10 and 100 and divide by 10.

Extension: During the photocopiable activity ask these learners to add extra columns for 100 000s and 1000ths. They multiply and divide by amounts that will use these columns.

Name: _____

Greater or less?

You will need:

A partner and a set of 0 to 9 digit cards.

What to do

- Pick three digit cards. Put them in a row to make a number and write the number in one of the tables below.

- In the row underneath multiply your number by 10, 100 or 1000.

- Then divide the original number by 10 or 100.

- Read all three numbers to your partner.

Here is an example:

10000s	1000s	100s	10s	1s	.	10ths	100ths
		6	3	8	.		
6	3	8	0	0	.		
			6	3	.	8	

10000s	1000s	100s	10s	1s	.	10ths	100ths
					.		
					.		
					.		

10000s	1000s	100s	10s	1s	.	10ths	100ths
					.		
					.		
					.		

10000s	1000s	100s	10s	1s	.	10ths	100ths
					.		
					.		
					.		

Comparing and ordering

Learning objectives

- Order and compare negative and positive numbers on a number line and temperature scale. (5Nn9)
- Investigate a simple general statement by finding examples which do or do not satisfy it, e.g. the sum of three consecutive whole numbers is always a multiple of three. (5Ps8)

Resources

Counting stick; pencils; paper; photocopiable page 75; scissors.

Starter

- Tell the learners that they will be practising counting in steps of different sizes. This may help in the lesson.
- Use the counting stick, with one end representing zero. Ask: *What size of steps will we count in to get from zero to 250?* Agree on 25. Together count in these steps from zero to 250 and back.
- Repeat for steps of 50, 100, 0.1, 0.2.

Main activities

- Write these numbers on the board: *23 645, 23 481, 23 654, 23 408, 23 451.* Ask the learners to order these numbers. Discuss which digits they had to look at to do this.
- Repeat with similar numbers.
- Ask the learners to make three-digit numbers for a partner to order.
- Invite pairs to share their orders. Choose two of their numbers and write them on the board. Ask: *What symbol can you put between the two to show comparison?* Establish that they could put > or < to show which is the larger and which the smaller.
- Write a set of six positive and negative numbers on the board, for example: *23, 14, –15, –3, 7, –10.* Ask the learners to tell you what –15 is.

- Agree that it is a number below zero. Ask: *Where might we see numbers below zero?* Agree temperatures in the colder parts of the world. Ask the learners to order the six numbers.
- Write this on the board: *–15 + 23 > * + *.* Ask the learners to use the numbers they ordered to find examples that can replace the stars. Write an example that can't and discuss why it won't fit.
- Repeat.
- Ask the learners to complete photocopiable page 75.

Plenary

- Ask the learners to describe to a partner what they have been learning about.
- Take feedback. Agree that they have been ordering and comparing positive and negative numbers.
- Discuss where negative numbers can be seen in real life.

Success criteria

Ask the learners:

- What does the symbol > mean? Can you give an example of when you would use it?
- What does the symbol < mean? Can you give an example of when you would use it?
- When do we use negative numbers in real life?

Ideas for differentiation

Support: During the photocopiable activity, these learners should make up four numbers each, and focus on numbers between –10 and 200.

Extension: Challenge these learners to use pairs of numbers to make up number sentences such as 145 + 45 > –23 + 152.

Name: _____

Ordering

You will need:

A partner, some paper, a pencil and scissors.

What to do

- Write down ten numbers. Make sure you have a variety with different numbers of digits. Make sure you include both positive and negative numbers.

- Cut out your numbers and pile them face down in front of you.

- With your partner, take it in turns to pick a number and place it face up on the table.

- As you pick numbers position them so that they are in order from lowest to highest.

- Write the order here:

_____

- Now choose pairs of numbers and write them with the symbols > or < to show comparison.

- Write these here:

_____

_____

_____

_____

_____

_____

_____

_____

_____

_____

_____

Decimal fractions 1

Learning objectives

- Use decimal notation for tenths and hundredths and understand what each digit represents. (5Nn4)
- Round a number with one or two decimal places to the nearest whole number. (5Nn7)
- Consider whether an answer is reasonable in the context of a problem. (5Pt7)

Resources

Set of follow-me cards from photocopiable page 11; pencils; paper; photocopiable page 77; set of digit cards from photocopiable page 9 for each learner.

Starter

- Tell the learners that they will be revising their times tables to keep them fresh in their minds.
- Share out the follow-me cards between small groups. Keep one card. Read out the multiplication, for example: *9 × 8*. The group with the answer calls it out and reads the multiplication on their card. Continue. You will have the last answer.
- Start with the answer on your card. The group that has the multiplication question that goes with it reads it out. They then read out the next answer and so on.

Main activities

- Write 3.7 on the board. Ask: *What can you tell me about this number? Discuss this with your partner.* Agree that it is a mixed number, three wholes and part of a whole. *The 3 is three ones and the 7 is seven tenths.* Establish that it could also be written as $3\frac{7}{10}$.
- Ask: *What would 3.7 be if we rounded it to the nearest whole number?* Agree 4 because 0.7 is closer to one than zero. Ask: *What about 3.5?* Agree that 3.5 also rounds up to 4.

- Ask the learners to write down some decimal numbers. They should also write them as a mixed fraction. Invite volunteers to share and explain their numbers. Together round each to the nearest whole number.
- Write 5.2 mm on the board. How else can we say this length? Agree 5 cm 2 mm or 52 mm. Ask: *What is the place value of each number?* Establish that the 5 represents whole centimetres and the 2 represents two tenths. Repeat with other lengths.
- Ask the learners to play the game on photocopiable page 77.

Plenary

- Take feedback from the photocopiable activity. Invite pairs of learners to share the numbers they made with the class and explain what each digit represents.
- Divide the class into two teams and play the game as in the photocopiable activity.

Success criteria

Ask the learners:

- What do the digits in 12.6 represent? How do you know?
- Which is higher, 15.7 or 15.70? How do you know?
- When do we see numbers with one decimal place in real life? When else?

Ideas for differentiation

Support: Ask these learners to make numbers with units and one decimal place only.

Extension: Ask these learners to make numbers with two decimal places.

Up the ladder!

You will need:

A partner and two sets of 0 to 9 digit cards.

What to do

- Shuffle the two sets of digit cards together and place them face down.

- Take it in turns to pick three cards.

- Use them to make a number with tens, units and one decimal place.

- Write the number on the ladder, with the smallest numbers at the bottom and the largest at the top.

- If you make a number that can't fit onto the ladder, you score 2 points.

- Put the cards at the bottom of the pile and pick three more.

- When you have filled the ladder, count your points.

- The player with the lowest score wins.

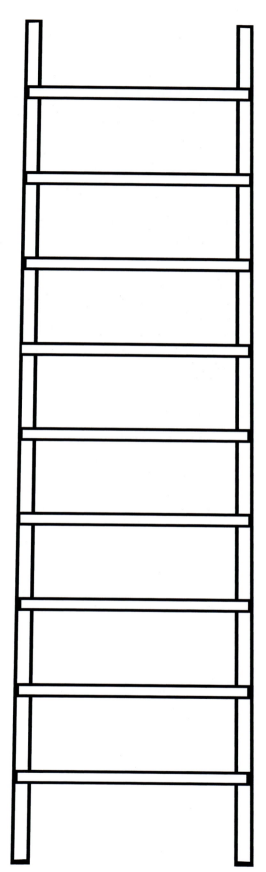

Decimal fractions 2

Learning objectives

● Round a number with one or two decimal places to the nearest whole number. (5Nn7)

● Order numbers with one or two decimal places and compare using the > and < signs. (5Nn11)

● Explain methods and justify reasoning orally; make hypotheses and test them out. (5Ps9)

Resources

Pencils; paper; photocopiable page 79; set of digit cards from photocopiable page 9 for each learner.

Starter

• Explain that the learners will be practising adding and subtracting pairs of two- and three-digit numbers.

• Ask the learners to give you pairs of two-digit numbers. The class then finds their totals and differences. Invite volunteers to share their strategies. Look out for strategies such as rounding and adjusting, near doubles, partitioning and sequencing.

• Repeat for three-digit numbers.

Main activities

• Write 23.71 on the board. Ask: *What can you tell me about this number? Discuss this with your partner.* Agree that the 2 represents two tens, the 3 three units, the 7 seven tenths and the 1 one hundredth. Establish that it could also be written as $23\frac{71}{100}$.

• Say: *I think that if I rounded 23.71 to the nearest whole number the answer would be 23. What do you think?* Agree that this would be wrong because 0.71 is above 0.5 so the answer would be 24.

• Say: *I think that if I rounded it to the nearest tenth it would be 23.7.* Discuss which digits they need to consider: 71 hundredths.

• Encourage the learners to use what they know to round. For example, establish that because 71 is closest to 70 then 71 hundredths would be closest to 70 hundredths. Agree that you are right.

• Repeat this with other numbers using 'I think…' statements.

• Ask the learners to write down some numbers with two decimal places. They should then pick pairs of these and compare them using the > and < symbols. Agree that > means 'is greater than' and < 'is less than'. Take feedback, inviting the learners to write their examples on the board.

• Ask the learners to work through the activity on photocopiable page 79.

Plenary

• Take feedback from the photocopiable activity. Invite individuals to share examples of the numbers they made, including how they rounded and then compared.

• As a class make a number with three decimal places. Round this to the nearest whole number, tenth and hundredth.

Success criteria

Ask the learners:

● 34.4 > 34.14. Is this correct? How do you know?

● Which is higher, 25.7 or 25.07? How do you know?

● When do we see numbers with two decimal places in real life? When else?

Ideas for differentiation

Support: Ask these learners to make numbers with one decimal place.

Extension: Ask these learners to make numbers with three decimal places.

Name: _____

Lots of decimals!

23.14

56.56

87.29

You will need:

A set of 0 to 9 digit cards.

What to do

- Pick four digit cards. Write them in the table.
- Make a number with two decimal places and write it in the table.
- Round it to the nearest whole number and then the nearest tenth.
- Use the same cards to make a second number and write it in the table.
- Round it to the nearest whole number and then the nearest tenth.
- Compare your two numbers using > or <.
- See how many numbers you can make.

| Digits picked | 1st number made | Rounded to | | 2nd number made | Rounded to | | Comparison |
		nearest whole number	nearest 10th		nearest whole number	nearest 10th	
3,6,9,1	36.19	36	36.2	91.36	91	91.4	36.19 < 91.36

Odd and even numbers

Learning objectives

- Make general statements about sums, differences and multiples of odd and even numbers. (5Nn14)
- Deduce new information from existing information to solve problems. (5Ps4)
- Use ordered lists and tables to help to solve problems systematically. (5Ps5)

Resources

Pencils; paper; photocopiable page 81; set of digit cards from photocopiable page 9 for each learner.

Starter

- Explain that, to help them during the lesson, the learners will revise odd and even numbers.
- Call out a mixture of odd and even numbers. When they hear you say an even number, the learners should put their hands in the air.
- Do this for a variety of three-, four-, five- and six-digit numbers.

Main activities

- Ask the learners to choose two even two-digit numbers and find their total. Ask: *Is the answer odd or even?* Agree that it is even. Ask them to choose two-different even numbers and repeat. They should then do this again with pairs of three- and four-digit even numbers. Model how making ordered lists to do this can help them work systematically.
- Ask: *What do you notice?* Agree that all the answers are even. Ask: *What is the general statement you can make about adding even numbers?*
- Divide the class into pairs and ask them to repeat the above with two-, then three- and four-digit odd numbers. Ask: *What do you notice about the totals this time?* Agree that they are also always even. Ask: *What general statement can you make about adding odd numbers?*

- Ask the learners to explore what they need to add to get an odd total. They should make lists and work systematically to do this. Take feedback, agreeing that an odd and an even number will give an odd total.
- Ask the learners to explore what happens when you subtract pairs of even or pairs of odd numbers. They should then explore subtracting with one of each.
- Take feedback. Agree that subtracting pairs of even numbers and odd numbers will give even results. Subtracting a mixture will give an odd result.
- Ask the learners to complete the activity on photocopiable page 81.
- Explore multiples of odd and even numbers in the same way.

Plenary

- Take feedback from the photocopiable activity. Invite pairs to share what they discovered.
- Discuss the idea that when adding a selection of odd and even numbers there needs to be an odd number of odd numbers for the answer to be odd.

Success criteria

Ask the learners:

- If you add 356 and 298 will you get an odd or even answer? How do you know without adding them up?
- Can you give me three numbers that will add together to give an odd answer? Are there any others?
- What generalisations can you make about adding and subtracting odd and even numbers?

Ideas for differentiation

Support: These learners should explore adding single- and two-digit numbers. Give them the opportunity to add larger numbers with a calculator.

Extension: During the photocopiable activity, ask these learners to make up four- and five-digit numbers.

Name: _____

Odd or even?

8, 56, 334

You will need:

A partner and a set of 0 to 9 digit cards.

What to do

- Pick three digit cards and make a three-digit number.

- Pick three more and make another.

- If you add them do you think the answer will be odd or even?

  Were you correct?

  _____

- Now pick another three digit cards to make a third number.

- If you add this to the other two numbers will your answer be odd or even?

  _____

  Were you correct?

  _____

- When you add two numbers you know that: E + E = E, O + O = E, O + E = O

1. What happens when you add three numbers?

    _____

2. What about when you add four numbers?

    _____

3. What have all the even answers got in common?

    _____

4. What have all the odd answers got in common?

    _____

5. What generalisations can you make about adding odd and even numbers?

    _____

Multiples and factors

Learning objectives

- Know and apply tests of divisibility by 2, 5, 10 and 100. (5Nc4)
- Recognise multiples of 6, 7, 8 and 9 up to the 10th multiple. (5Nc5)
- Find factors of two-digit numbers. (5Nc7)
- Explore and solve number problems and puzzles, e.g. logic problems. (5Ps3)

Resources

Pencils; paper; photocopiable page 83.

Starter

- Explain that the learners will revise the tests of divisibility for 2, 5, 10 and 100.
- Quickly recap what they are: 2 (even number), 5 (ends in 5 or 0), 10 (ends in 0), 100 (ends in 00). Call out a variety of numbers, for example: *24, 75, 160, 3400.* The learners write down the numbers they are divisible by using their knowledge of these divisibility tests.

Main activities

- Ask: *What is a multiple?* Agree that it is a number made by multiplying together two other numbers. For example, 32 is a multiple of 4 and 8 because 8 × 4 = 32. Ask the learners to write down as many multiples of 6 as they can in a minute.
- Take feedback. Focus on larger multiples. Discuss how they can double and multiply by 10 or 100 to find them, for example, a large multiple of 72 is 1440 (double and multiply by 10). Give them another minute to add to their list by doing this or using other strategies that they can think of. Invite individuals to share some of their multiples.
- Ask: *What is a factor?* Agree that it is a number that can be divided into another without a remainder.

- Ask the learners to write the factors of various two- and three-digit numbers, for example 24, 60, 100. Take feedback and list them on the board.
- Point out the numbers that are factors of more than one of those numbers. Agree that some numbers can be factors of several other numbers.
- Ask the learners to work in groups of three or four to complete the puzzle on photocopiable page 83.

Plenary

- Take feedback on the photocopiable activity. Discuss which numbers went where in the grid and why.
- Ask the learners who made up their own grid to share it with the class.
- Note that there isn't one unique solution to this activity.
- Finish by recapping what multiples and factors are.

Success criteria

Ask the learners:

- How would you explain what a multiple is to someone who didn't know? What about a factor? Is there another way to explain?
- Can you give us a number that is a factor of both 24 and 32? Are there any others?
- Can you give us a four-digit number that is a multiple of 7? How do you know?

Ideas for differentiation

Support: During the main activity, ask these learners to work on two-digit numbers; during the photocopiable activity they should focus on the first two rows of the table.

Extension: Challenge these learners to find three- and four-digit multiples.

Name: _____

Multiples or factors?

1.   Work in a group of three or four.

2.   The numbers below need to be placed in the correct position in the table.
     Only one number can go into each section!

| 60 | | 21 | | 1 | | 40 | | 9 | | 6 | | 8 | | 50 |
| 48 | | 4 | | 30 | | 45 | | 2 | | 18 | | 3 | | 12 |

	Multiple of 3	Factor of 24	Multiple of 5	Factor of 36
Even number				
Odd number				
Multiple of 6				
Factor of 120				

3.   Think of other numbers to add to each section.

4.   Now make up a grid of your own. You can choose which multiples and factors to
     use. Draw your grid here.

Mental calculation strategies

Learning objectives

- Describe and continue number sequences, e.g. −30, −27, *, *, −18, …; identify the relationships between numbers. (5Ps6)
- Count on or back in thousands, hundreds, tens and ones to add or subtract. (5Nc8)
- Add or subtract near multiples of 10 or 100, e.g. 4387 − 299. (5Nc9)
- Calculate differences between near multiples of 1000, e.g. 5026 − 4998, or near multiples of 1, e.g. 3.2 − 2.6. (5Nc11)

Resources

Counting stick; pencils; paper for recording; photocopiable page 85; set of 0 to 9 digit cards from photocopiable page 9 for each learner.

Starter

- Explain to the learners that they will rehearse describing and continuing number sequences.
- Show the counting stick. Tell the learners that 24 is at the end on the right. Point to the next three divisions and say: *29, 34, 39*. The learners should work out the sequence and continue counting to the other end.
- Repeat for other forward and backward sequences, for example: *25, 16, 7, −2*. Start at the end on the left for counting backwards.

Main activities

- Set this problem: *Ella had 249 shells. Sofia had 199. How many did they have altogether? How many more did Ella have?* Discuss what to do to answer the problem: add for the total, subtract for finding the difference.
- Ask the learners to think of different ways to answer using mental calculation strategies. Take feedback, inviting individuals to demonstrate their methods.
- For the total, model these methods:
  - add one hundred to 249, then 90 and 9 (sequencing)
  - add 200 and take away one (rounding and adjusting)

- For the difference, model these methods:
  - take away 100, then 90 and 9 (sequencing)
  - take away 200, then add one (rounding and adjusting)
  - add 1 to make 200 and then 49 (counting up)
- Discuss which are the most efficient methods to add and subtract these numbers and why.
- Write three- and four-digit calculations on the board. Ask the learners to practise adding and subtracting these using the methods discussed.
- Ask the learners to complete photocopiable page 85.

Plenary

- Invite the learners to share the calculations they made up and how they used each of the methods to solve them.
- Ask the learners to evaluate each method and decide which they thought was the most efficient for each calculation.
- Finish the lesson by asking the learners to tell you all the methods that they now know for addition and subtraction.

Success criteria

Ask the learners:

- What is the most efficient way to add 4589 and 2998? Why do you think this is the most efficient?
- What is the most efficient way to subtract 1900 from 2576? Why do you think this is the most efficient? Is there another efficient method?
- Can you make up a calculation that would be answered best by sequencing?

Ideas for differentiation

Support: Ask these learners to focus on adding and subtracting two- and three-digit numbers.

Extension: Ask these learners to focus on adding and subtracting four- and five-digit numbers.

Name: _____

Efficient methods!

You will need:

A set of 0 to 9 digit cards.

What to do

- Shuffle the digit cards. Take the top six and use them to make two three-digit numbers.

- Find their total using the methods discussed in the lesson.

- Find their difference using the methods discussed in the lesson.

- Record your work in the grid below.

- Do this again with another six cards.

- Now take eight cards and do the same with two four-digit numbers.

An example has been done for you.

Numbers made	Addition calculations		Subtraction calculations		
	Sequencing method	Rounding and adjusting	Sequencing method	Rounding and adjusting	Counting on
345, 628	628 + 300 + 40 + 5 = 973	345 + 630 – 2 = 973	628 – 300 – 40 – 5 = 283	628 – 350 + 5 = 283	345 + 55 + 228 = 628, 55 + 228 = 283

- Make up a problem for each set of calculations. Write them on the back of this paper.

Addition and subtraction 1

Learning objectives

- Estimate and approximate when calculating, e.g. using rounding, and check working. (5Pt6)
- Check with a different order when adding several numbers or by using the inverse when adding or subtracting a pair of numbers. (5Pt3)
- Solve a larger problem by breaking it down into sub-problems or represent it using diagrams. (5Ps10)

Resources

Pencils; paper; photocopiable page 87.

Starter

- Tell the learners that they will be adding and subtracting near multiples of 100.
- Write these calculations on the board: *342 + 199, 143 + 98, 354 − 199, 132 − 97*. The learners solve these by rounding and adjusting and then write down their answers.
- Repeat with similar calculations.

Main activities

- Recap the mental calculation strategies for addition from page 18.
- Ask the learners to make up calculations that could be solved using each of these methods. Invite volunteers to share examples, which the rest of the class then solve.
- Discuss how to check that answers are correct. Establish that they can add in a different order or subtract one of the numbers from the answer (inverse). Ask the learners to do this for their answers to the calculations.
- Discuss possible written methods for calculations that cannot easily be solved mentally. Demonstrate the following:
  - Partitioning (most significant digits first):

	2	3	5	6
+	1	4	4	6
	3	0	0	0
		7	0	0
			9	0
			1	2
	3	8	0	2

- Compact method, if appropriate:

+	2	4	5	3
+	1	9	2	8
	4	3	8	1
	1		1	

- Write calculations on the board for the learners to solve in one of these ways.
- Set this problem: *Leona has saved $50. She wants to buy a music player for $23.48. She also wants to download music from the internet. This will cost $9.67. Does she have enough money left to buy some headphones at $8.96?*
- Ask the learners to estimate the answer first by rounding, then work with a partner to solve the problem. Take feedback, discussing all the methods used.
- Ask the learners to work through photocopiable page 87.

Plenary

- Write some calculations on the board. Ask the learners to estimate the answers first and then choose a strategy to use to solve them.

Success criteria

Ask the learners:

- What mental calculation strategies could we use for adding 245 + 249?
- What type of calculation could be solved by adding a multiple of 10 and adjusting?
- How can you solve 3425 + 2973? Is there another way?

Ideas for differentiation

Support: During the lesson, ask these learners to work with two-digit numbers. For the photocopiable activity, ask them to total two or three items.

Extension: During the photocopiable activity, ask these learners to use their preferred written method for finding the total, and check their answer using another.

Name: _____

Addition or subtraction?

1. A company is making a list of food they would like to buy for their office party.

   This is their list so far with the prices for the amounts they need:

Food	Price
French stick	$4.45
Doughnuts	$9.99
Ice cream	$15.25
Pizza	$42.80
Samosa	$4.50
Aloo tikka	$10.75

2. They have $75. How much more money do they need to buy everything on this list? Estimate first. Write your estimate and any jottings here.

   _____

   _____

   _____

3. You can use any strategies you want for your calculations. Show your workings in the space below.

Addition and subtraction 2

Learning objectives

● Use appropriate strategies to add or subtract pairs of two- and three-digit numbers and numbers with one decimal place, using jottings where necessary. (5Nc10)

● Add or subtract any pair of three- and / or four-digit numbers, with the same number of decimal places, including amounts of money. (5Nc19)

● Solve single and multi-step word problems (all four operations): represent them, e.g. with diagrams or on a number line. (5Pt2)

● Choose an appropriate strategy for a calculation and explain how they worked out the answer. (5Ps2)

Resources

Pencils; paper; photocopiable page 89.

Starter

• Explain that the learners will practise calculating differences between near multiples of 1000 and 1.

• Write 2134 – 1998 on the board. Ask the learners to find the difference by counting up from 1998 to 2134 and write down the answer.

• Repeat for other two-, three- and four-digit numbers and decimals, for example 2.4 – 1.9. Direct learners of different attainment levels to the appropriate calculations.

Main activities

• Recap the mental calculation strategies for subtraction from page 20.

• Ask the learners to make up calculations that could be best solved using each of these methods. Invite volunteers to share examples, which the rest of the class then solves using the strategy that the volunteer suggests.

• Ask: *If we have to solve a calculation that cannot easily be solved using one of these strategies, what could we do?* Agree that a written method might be more appropriate.

• Recap these methods for subtraction:
  • Complementary addition:

    $$3425 - 2674 \rightarrow 2674 + 26 = 2700, 2700 + 300 = 3000, 3000 + 425 = 3425$$
    $$\rightarrow 26 + 300 + 425 = 751$$

  • Sequencing:

    $$4256 - 2931 \rightarrow 4256 - 2000 = 2256, 2256 - 900 = 1356, 1356 - 30 = 1326, 1326 - 1 = 1325$$

  • Decomposition, if appropriate:

	4	$\overset{3}{\cancel{4}}$^{2}	$\overset{1}{\cancel{2}}$^{1}	^{1}7
–	2	2	4	9
	2	0	7	8

• Write calculations on the board for the learners to solve in one of these ways.

• Set this problem: *Ralph spent $24.99 on a pair of jeans. He gave the store assistant $50. How much change will he receive?*

• Ask the learners to estimate the answer first by rounding, then work with a partner to solve the problem. Take feedback, discussing methods.

• Ask the learners to work through photocopiable page 89.

Plenary

• Write some calculations on the boards for the learners to answer using their preferred method.

Success criteria

Ask the learners:

● How can you explain sequencing to someone who doesn't know? Is there another way?

● How would you calculate 256 – 179? Is there another way?

● Can you give us a calculation that can be answered using complementary addition?

Ideas for differentiation

Support: During the photocopiable activity, these learners should work on the first three shops only and focus on complementary addition.

Extension: During all activities, ask these learners to use their preferred method and then check their results with a mental calculation strategy.

Name: _____

How much is left?

Waleed wanted to buy some packs of football cards with his pocket money.

He had saved $20. He made a note of how much the cards cost at different shops.

1.   Work out how much he will have left if he buys the cards from each shop.

2.   Use the space in the table to show your working.

Shop	Price	Change from $20 Show your working
Cards4u	$10.50	
Card City	$9.45	
Football Mania	$11.15	
Footie Stuff	$8.93	
Teams United	$12.01	
Sport4all	$10.79	

3.   If he wanted two packs, which shops could he buy them from?

_____

_____

Multiplication 1

Learning objectives

- Multiply by 25 by multiplying by 100 and dividing by 4. (5Nc14)
- Multiply or divide three-digit numbers by single-digit numbers. (5Nc20)
- Solve single and multi-step word problems (all four operations); represent them, e.g. with diagrams or a number line. (5Pt2)

Resources

Set of follow-me cards from photocopiable page 11; pencils; paper; photocopiable page 91; set of 0 to 9 digit cards from photocopiable page 9 for each learner.

Starter

- Share out the follow-me cards between small groups. Keep one card. Read out the multiplication, for example: *4 × 6*. The group with the answer calls it out and reads the multiplication on their card. Continue until all cards have been read.
- Start with the answer on your card. The group that has the multiplication question that goes with it reads it out, then reads out the next answer, and so on.

Main activities

- Set this problem: *Cherri, a librarian, has 12 boxes of books to put on the shelves in the library. In each box there are 25 books. How many books will she unpack?* Ask the learners to discuss in pairs how they could find out.
- Take feedback. Agree that they need to multiply the numbers together. Discuss possible strategies. Focus on the mental method of multiplying by 25 by first multiplying by 100 and then dividing by 4 (1200 ÷ 4 = 300).
- Ask: *If we are multiplying by a number that doesn't encourage a mental method, what could we do?* Agree that the grid method might help.

- Ask a volunteer to demonstrate for 245 × 8:

	200	40	5
8	1600	320	40

1	6	0	0
	3	2	0
+		4	0
1	9	6	0

- Encourage the learners to explain that, for example, because they know that 4 × 8 is 32, they know that 40 × 8 is 320.
- Write calculations, similar to the example above, on the board so the learners can practise.
- Repeat the Cherri scenario for different sets of numbers or make up your own.
- Ask the learners to complete photocopiable page 91.

Plenary

- Invite volunteers to share the problems they made up in the photocopiable activity. Find the solutions as a class. Each time, discuss the best method to use.
- Ask the learners to assess their confidence in using mental calculation methods and the grid method.

Success criteria

Ask the learners:

- What is 146 multiplied by 4? How did you work that out? Is there another way?
- If we know that 8 × 9 is 72, what else do we know? Is there anything else?
- How would you explain how to multiply decimal numbers using the grid method to someone who didn't know?

Ideas for differentiation

Support: During the photocopiable activity, ask these learners to focus on multiplying two-digit numbers and one-digit numbers with one decimal place.

Extension: Challenge these learners to multiply four-digit numbers and two-digit numbers with two decimal places.

Name: _____

Multiplication problems

You will need:

A set of 0 to 9 digit cards.

What to do

- You are going to make up six multiplication calculations.

- Take three digit cards and use them to make a three-digit number, for example 452.

- Pick a fourth digit card. Multiply your three-digit number by this number.

- You can use a mental method or the grid method.

  Here are two examples:

  312 × 4: double and double again → 624, 1248

  596 × 8: grid method →

	500	90	6	
8	4000	720	48	Total: 4768

- Show your working in the table below.

- Do this two more times.

- Do the same again three more times, but now use the three digit cards to make a two-digit number with one decimal place, for example 62.3

Numbers picked	Calculation	Solution

- Choose three of your calculations. Make up a problem to go with them. Write your problems on the back of this paper.

Multiplication 2

- Multiply multiples of 10 to 90, and multiples of 100 to 900, by a single-digit number. (5Nc12)
- Multiply by 19 or 21 by multiplying by 20 and adjusting. (5Nc13)
- Use factors to multiply, e.g. multiply by 3, then double to multiply by 6. (5Nc15)
- Multiply two-digit numbers by two-digit numbers. (5Nc21)

Resources

Pencils; paper; photocopiable page 93; dice; counters; calculators.

Starter

- Explain that the learners will practise multiplying multiples of 10 and 100 by single-digit numbers.
- Call out multiples of 10 and ask the learners to multiply these by 6, 7 and 8. Invite the learners to share their strategies, for example $9 \times 8 = 72$, so $90 \times 8 = 720$.
- Repeat for multiples of 100 to 900.

Main activities

- Write this calculation on the board: *45 × 19*. Ask the learners to suggest ways of solving this. Take feedback and discuss the idea of multiplying by 20 and taking one lot of 45 away: $900 - 45 = 855$. Write similar calculations on the board for the learners to answer using this method.
- Repeat for multiplying by 21, this time adding instead of taking away.
- Focus on using factors to multiply. Write this on the board: *64 × 6*. Agree that you can multiply by 3 and double: $192 \times 2 = 384$. Repeat using similar calculations.
- Focus on multiplying two-digit by two-digit numbers that could be answered using a written method. Set this problem: *Georgie has 24 packets of sweets. In each packet are 36 sweets. How many sweets does she have altogether?*

- Invite individuals to share their ideas on how to answer the problem. Agree that they could use the grid method:

	20	4
30	600	120
6	120	24

Find the total by adding the hundreds, then the tens and finally the ones: $800 + 60 + 4 = 864$

- Ask the learners to work in pairs to make up a problem to ask the class. This should involve two-digit by two-digit multiplication.
- Invite pairs to share their problems. Solve them as a class.
- Ask the learners to play the game on photocopiable page 93.

Plenary

- Invite pairs of learners to demonstrate how they solved some of the calculations they made up during the game.
- Ask the learners to assess how confident they are at multiplying using the grid method.

Success criteria

Ask the learners:

- How would you describe how to solve 12.4×7 to someone who doesn't know? Is there another way?
- What is 76×48? How did you work that out?
- Can you make up a problem that involves multiplying 36 by 14?

Ideas for differentiation

Support: Ask these learners to focus on multiplying two-digit numbers by a single-digit number.

Extension: After playing the game on the photocopiable page, challenge these learners to make up their own version with three-digit numbers and numbers with two decimal places.

Multiplication game

You will need:

A partner, a dice, two counters and a calculator.

What to do

● Take it in turns to roll the dice.

● Move your counter that number of squares.

● Choose a two-digit number to multiply the number on the square by (not a multiple of 10!)

● Check your answer with a calculator.

● Stay on this square if you are correct. If your answer is wrong move your counter back to where it was before.

● The first person to the end wins.

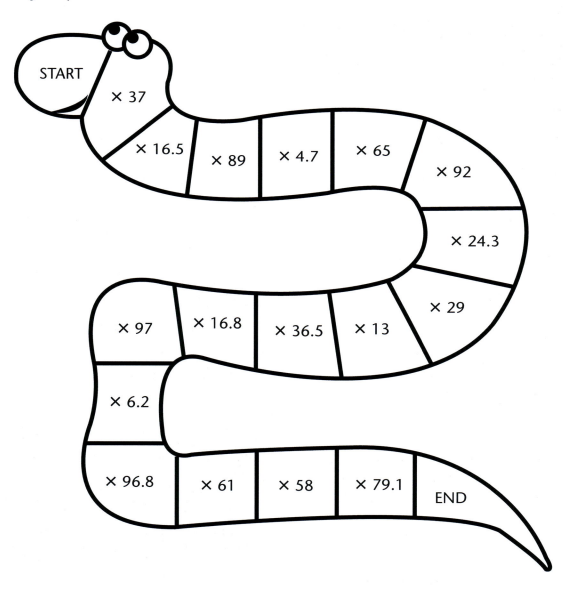

START
× 37
× 16.5
× 89
× 4.7
× 65
× 92
× 24.3
× 29
× 13
× 36.5
× 16.8
× 97
× 6.2
× 96.8
× 61
× 58
× 79.1
END

Cambridge Primary: Ready to Go Lessons for Maths Stage 5 © Hodder & Stoughton Ltd 2012

Division

- Double any number up to 100 and halve even numbers to 200 and use this to double and halve numbers with one or two decimal places, e.g. double 3.4 and half of 8.6. (5Nc16)

- Double multiples of 10 to 1000 and multiples of 100 to 10 000, e.g. double 360 or double 3600, and derive corresponding halves. (5Nc17)

- Divide three-digit numbers by single-digit numbers, including those with a remainder (answers no greater than 30). (5Nc23)

- Use multiplication to check the result of a division, e.g. multiply 3.7 × 8 to check 29.6 ÷ 8. (5Pt4)

Resources

Pencils; paper; photocopiable page 95; scissors.

Starter

- Explain to the learners that they will revise doubling and halving.

- Call out some numbers, for example: *72, 35.* Ask the learners to double these. Ask: *How can we use our knowledge of these to double 7.2 and 3.5?* Agree that the answer will be similar, just 10 times smaller.

- Repeat for halving numbers such as 124 and 12.4.

- Ask the learners to double multiples of 10 and 100, for example 480, 4800. Then ask them to halve these doubled numbers (960, 9600).

Main activities

- Set this problem: *Auzma made 178 cakes. She wants to put them in boxes. She can only put 8 in each box. How many boxes does she need?* Ask the learners to discuss in pairs what they need to do to solve this.

- Establish that they need to take away multiples of the divisor. Ask the learners to do this, taking away the largest multiples they can.

- Take feedback. Invite volunteers to demonstrate their methods, for example:

	1	7	8	÷ 8
−	1	6	0	(2 lots of 8 is 16, so 20 lots of 8 is 160)
		1	8	
−		1	6	(2 lots of 8 is 16)
			2	

- Agree that Auzma will need 22 boxes for 176 cakes. Discuss what she could do with what is left. For example, she could eat them, which will mean the answer is rounded down, or she could put them in another box, in which case the answer will be rounded up.

- Discuss how the learners can check this. Establish that they can multiply 22 by 8 and add 2.

- Write some calculations on the board so that the learners can practise. After each calculation they should check by multiplying.

- Ask similar problems to Auzma's, each time discussing what to do with the remainder.

- Ask the learners to complete photocopiable page 95 in groups of four.

Plenary

- Take feedback from the photocopiable activity, inviting the learners to share their methods.

Success criteria

Ask the learners:

- What is meant by grouping? How could you explain how to solve 356 ÷ 6? Is there another way?

- Can you make up a problem that involves grouping?

- How would you know whether to round up or down after a division calculation?

Ideas for differentiation

Support: Adapt photocopiable page 95 so these learners divide two-digit numbers by a single digit.

Extension: Adapt page 95 so these learners divide three-digit numbers by two-digit numbers.

Solve that problem!

You will need:

Scissors.

What to do

- Work in a group of four. Organise your group into two pairs.
- Cut up the problems and deal them out between the two pairs.
- Answer each problem with your partner.
- When you have answered them, explain what you did to the other pair.

---

1. Books need to be stacked in a storage box in the library.

   Each shelf holds 9 books. There are 145 books.

   a) How many boxes are completely filled?

   b) How many boxes are needed altogether?

---

2. My little sister had a total of 117 hours sleep on her 9-day trip to France.

   How many hours sleep is this per day?

---

3. The truck container bay is 150 m long. Each container is 2.5 m long.

   How many containers can be fitted in lengthways?

---

4. Sandy has a sheet of 250 stamps. He tears them into strips of 8.

   a) How many strips does he have?

   b) How many stamps are left over?

---

5. The baker had baked 200 loaves of bread. He packed them in containers.

   He wanted to put 6 loaves in each one.

   How many containers does he need for all his loaves?

---

6. Esme counted out the money in her money box. She had $45.60.
   She decided to put it into bags. She put $7 in each bag.

   a) How many full bags did she have?

   b) How much was left over?

Unit assessment

- What is the value of each digit in 254 096?
- Explain what happens when you multiply a number by 100.
- Why is 245 divided by one hundred 2.45?

- What is the value of the 4 in 26.4?
- Explain how you know that 240 is a multiple of 2, 10 and 100.

Summative assessment activities

Observe the learners while they take part in these activities. You will quickly be able to identify those who appear to be confident and those who may need additional support.

Place value

This activity assesses the learners' knowledge of place value.

You will need:

Set of 0 to 9 digit cards made from photocopiable page 9; paper; pencils.

What to do

- Organise the learners into groups of four.
- Shuffle the digit cards.
- Ask the learners to take it in turns to pick a card and lay it on the table.
- Each new card should be placed to the left of the previous one.
- Each time a card is laid, ask the learners to read the number.
- Stop when you have a six-digit number.
- Point to different cards.
- Ask the learners to write down the value of the card you point at, for example for 348 195, point to the 8; the learners write down 8000.

Factors

This activity assesses the learners' knowledge of factors and multiples.

You will need:

Paper; pencils; stop watch.

What to do

- Organise the learners into groups of three or four.
- Write the number 36 on a piece of paper and show it to the learners.
- Time them for a minute. During that minute they write down all the factors of 36.
- Write the number 6 on another piece of paper.
- Time them for a minute. This time they write down as many multiples of 6 as they can in that minute.

Distribute copies of photocopiable page 97. Ask the learners to read the questions and write the answers. They should work independently.

Name: _____

Working with numbers 2

1. Write down the values of the digits in this number: 592 810

    _____

2. Why does this number need a zero? 245 609

    _____

3. Write down what you know about the multiples of 2, 5 and 10:

    2: _____

    5: _____

    10: _____

4. Draw a circle around all the multiples of 9:

    18   29   36   49   63   94

5. Draw a circle around the factors of 72:

    3   4   5   6   7   8

6. Raj bought a CD for $12.99 and a DVD for $20.75.

    Estimate how much she spent.

    _____

    Now find out how close your estimate was.

    _____

7. Arjun had $50. He spent $25 on trainers and $13.75 on a T shirt.

    How much money did he have left?

    _____

8. Sadeek had 198 one-cent coins.

    He put them into bags. He wanted to put 5 cents in each bag.

    How many bags did he need?

    _____

Unit 2B: Data handling and problem solving

Frequency tables

Learning objectives

- Draw and interpret frequency tables, pictograms and bar line charts, with the vertical axis labelled for example in twos, fives, tens, twenties or hundreds. Consider the effect of changing the scale on the vertical axis. (5Dh2)
- Deduce new information from existing information to solve problems. (5Ps4)
- Use ordered lists and tables to help solve problems systematically. (5Ps5)

Resources

Counting stick; pencils; plain paper; rulers; photocopiable page 99; dice.

Starter

- Explain that the learners will count on and back in steps of different sizes.
- Point to the middle division on the counting stick. Ask: *If this is zero, what numbers will go at either end?*
- Agree on −40 and 40. Count in steps from zero to 40 and then back to −40.
- Repeat this for other numbers.

Main activities

- Ask: *What is a frequency table?* Establish that it is a table that shows information about how often things happen. Give this example: *Ten learners took a test; one scored 100, 2 scored 90, six 80 and one 70.* Draw this on the board to illustrate:

Score	Number of learners
100	1
90	2
80	6
70	1

- Use it to explain that a frequency table shows how often a score was achieved, for example 80 was scored six times.

- Say: *I think apples are the most popular fruit in this class. What do you think? How can we find out?* Ask the learners to discuss this in pairs.
- Take feedback. Agree that everyone can be asked what their favourite fruit is. This information can be placed in a frequency table. Draw one on the board. List six fruits, including apples, in the left-hand column. Take a vote to find out which fruit is each learner's favourite. Write this in the right-hand column. Discuss whether your statement was correct.
- Invite the learners to make up a statement. They take a vote and draw their own frequency table on plain paper to show the results and find out if the statement is correct.
- Ask the learners to work through photocopiable page 99.

Plenary

- Ask the learners to share their findings from the photocopiable activity.
- Ask them to share the statements they made and describe how they found out if they were correct or not.

Success criteria

Ask the learners:

- How would you describe a frequency table to someone who doesn't know?
- What information does it give us? Can you give an example?
- How could we use one to prove or disprove a statement? Is there anything else we need to do?

Ideas for differentiation

Support: Give these learners an empty frequency table to use during the photocopiable activity.

Extension: During the photocopiable activity, give these learners an opportunity to explain the activity to a learner who needs some help.

Name: _____

Frequency table

1. Find out if this statement is true:

   Everyone in this class likes football.

2. What do you think about the statement? How can you find out if it's true?

3. Use this space to draw a frequency table and prove whether the statement is true or not.

4. Draw a frequency table to find out whether this statement is true or not:

   Every time you throw a dice it will land on an even number.

5. Now make up a statement of your own. Use a frequency table to find out if you are correct.

   _____

Pictograms

- Answer a set of related questions by collecting, selecting and organising relevant data; draw conclusions from their own and others' data and identify further questions to ask. (5Dh1)
- Deduce new information from existing information to solve problems. (5Ps4)
- Explain methods and justify reasoning orally and in writing; make hypotheses and test them out. (5Ps9)

Counting stick; pencils; plain paper; photocopiable page 101.

Starter

- Explain that the learners will count on and back in steps of different sizes.
- Point to one end of the counting stick. Say: *This is zero, the other end is 1. What steps will we count in to get from zero to one?*
- Agree on 0.1. Count in steps from zero to 1 and back. Put your finger on different divisions. Ask: *What number goes here?*
- Repeat with other decimal numbers.

Main activities

- Say: *I think potato chips are the most popular snack in this class. How can we find out?* Ask the learners to discuss this in pairs, then take feedback.
- Agree on seven possible snacks, including potato chips, and take a vote. Display the information in a frequency table. Ask: *How else can we display this information?* Take feedback, listing their ideas on the board.
- Focus on pictograms: representations using symbols.
- Write the seven snacks in a line on the board. Invite the learners to draw a smiley face above the snack they chose.

- Ask: *How can we create a pictogram with fewer faces?* Establish that each face could represent two learners. Adapt your pictogram.
- Ask questions such as: *Are chips the most popular snack in the class? How many of you prefer (snack) to (snack)?*
- Ask the learners to decide on something to find out from the class, for example hobbies. Together collect the data. Ask the learners to create their own pictogram with symbols representing two learners.
- Ask the learners to work through photocopiable page 101.

Plenary

- Invite the learners to share how they solved the problem from the photocopiable activity.
- Discuss the different symbol representations and how they affect the appearance of the pictogram.
- Ask questions based on their pictograms. Invite them to ask questions.

Ask the learners:

- How would you explain how to create a pictogram?
- What can you tell me about the symbols on a pictogram? What else?
- How are frequency tables and pictograms the same? How are they different?
- Which is the least popular snack? How do you know?

Support: During the photocopiable activity, allow these learners to use each symbol to represent one person.

Extension: During the photocopiable activity, challenge these learners to use each symbol to represent four people.

Pictograms

Noah wants to open a pet shop. He wants to begin by selling pets from his home. Because his house is small he wants to stock just six pets. He wants you to give him an idea of which animals to sell.

1.   How could you help him?

2.   Use the ideas of all the other learners in the class.

3.   Draw a pictogram to show your results.
     Don't forget to label your pictogram and
     say what the symbols represent.

4.   What advice would you give Noah?

_____

_____

_____

_____

Bar line charts

Resources

Pencils; plain and squared paper; rulers; photocopiable page 103; books.

Starter

- Explain that the learners will count in 1000s, 100s, 10s and 1s to keep this skill fresh in their minds.
- Together count in these steps from different numbers, for example 5698.
- Write some additions on the board. Ask the learners to find the totals by partitioning the second number, adding each part separately.

Main activities

- Ask: *In what ways can we represent data? Talk to a partner about this.* Take feedback. Agree that they could use, for example, frequency tables and pictograms. Invite the learners to demonstrate what both of these look like on the board (see pages 98 and 100 for examples).
- Focus on bar line charts. Ask: *What do you think a bar line chart looks like?* Invite volunteers to demonstrate what they think on the board.
- Draw this example on the board:

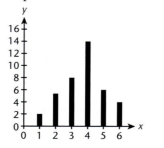

Discuss what it tells us. Look at the vertical axis and discuss the intervals. Ask the learners to work with a partner to make up what this chart might be about. Take feedback. Choose a scenario and add the appropriate labels.

- Set this scenario: *Fabio wanted a new kit for his football team. He can't decide between red, blue, yellow, green, purple and black. Can you decide?* Take a class vote. Ask the learners to work in pairs to make a bar line chart showing the information. Remind them to consider the intervals on the vertical axis and the labels needed.
- Invite pairs to share what they did.
- Divide the class into mixed-ability groups of four and ask them to work through photocopiable page 103.

Plenary

- Invite groups to share their results from the photocopiable activity.
- Ask: *Which was the most frequent vowel?* The learners should make a prediction first; then take feedback from all the groups.
- Ask the learners to assess their confidence in constructing bar line charts.

Success criteria

Ask the learners:

- Can you explain how to construct a bar line chart to someone who doesn't know? Can you explain in a different way?
- Can you tell us something about the numbers on the vertical axis?
- When are charts like this used in real life? When else?

Ideas for differentiation

Support: During the photocopiable activity, allow these learners to draw their bar line charts on squared paper.

Extension: Once these learners have completed the photocopiable activity, challenge them to think of something they would like to find out, collect the data from the class and create a bar line chart.

Name: _____

Bar line charts

1. Work in a small group.

2. Choose a book.

3. Take one page and predict which vowel will occur most often and which vowel will occur least often.

   Vowels are the letters a, e, i, o and u.

4. Test out your prediction. Count all the vowels you can see. Make a tally for each one.

5. Represent the results in a bar line chart.

6. Draw your chart on squared paper.

7. Compare your findings with another group.

_____

_____

_____

Line graphs 1

Learning objectives

- Construct simple line graphs, e.g. to show changes in temperature over time. (5Dh3)
- Understand where intermediate points have and do not have meaning, e.g. comparing a line graph of temperature against time with a graph of class attendance for each day of the week. (5Dh4)
- Deduce new information from existing information to solve problems. (5Ps4)

Resources

Pencils; paper; photocopiable page 105.

Starter

- Explain that the learners will rehearse counting in steps of different sizes.
- Ask the learners to write down all the multiples of 9 to the 10th multiple.
- Invite individuals to call out one of their multiples. The rest of the class write down the multiplication number sentence to go with it.
- Repeat for multiples of 6, 7 and 8.

Main activities

- Ask: *How can we represent information?* Agree, for example, frequency tables, pictograms and bar line charts. Invite volunteers to sketch these on the board.
- Explain that, today, they will be learning about line graphs. Sketch this example on the board:

People in a shop

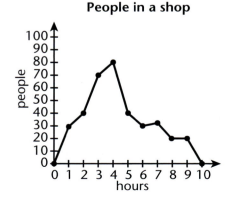

Ask the learners to explain what this shows: the number of people shopping over a period of 10 hours. Discuss the idea of discrete and continuous data: discrete data is the result of counting something at a given period of time, continuous data is something that is happening over a period of time. Discuss the intervals on the vertical axis.

- Ask the learners to estimate how many people were in the shop at different hourly points. Discuss the points between the hours. Establish that there are still people shopping. Agree that this is continuous data because something is happening all the time during the 10 hour period.
- Ask the learners to work in a small group and make up another scenario for the line graph. Take feedback.
- Ask the learners to work through photocopiable page 105 with a partner.

Plenary

- Invite groups to share the stories they made up in the photocopiable activity.
- Ask the class to ask each group questions to answer from their graph.
- Discuss strategies for finding the total distance travelled.

Success criteria

Ask the learners:

- Can you describe how to construct a line graph? Is there another way?
- What is meant by discrete data? Can you give examples?
- What is meant by continuous data? Can you give examples?
- When do we see line graphs in real life?

Ideas for differentiation

Support: Ensure these learners are supported by working in mixed-ability groups.

Extension: These learners should take the lead in their groups, encouraging everyone else to take part.

Name: _____

The story of a line graph

1. Work with a partner.

2. Look carefully at the line graph below.

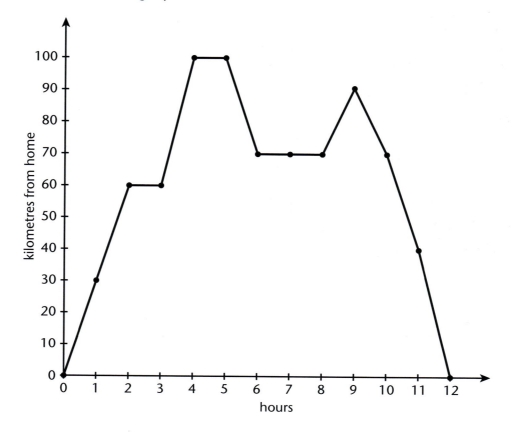

3. Make up a story about it. Make sure you talk about what is happening when there is no movement.

4. Write your story here:

_____

_____

_____

_____

_____

_____

_____

Line graphs 2

Learning objectives

- Calculate a rise or fall in temperature. (5Nn10)
- Construct simple line graphs, e.g. to show changes in temperature over time. (5Dh3)
- Find and interpret the mode of a set of data. (5Dh5)

Resources

Counting stick; pencils; paper for recording; photocopiable page 107; squared paper.

Starter

- Explain that the learners will think about temperature differences.
- Hold the counting stick vertically. Ask the learners to imagine it is the scale on a thermometer. Make the middle division zero. The intervals are to be ten.
- Point to an interval and ask what number would go there. Repeat for another. Ask: *The temperature has risen from here to here. How much has it risen?*
- Repeat for other pairs of 'temperatures'. Include estimating for positions between intervals.

Main activities

- Ask: *What is a line graph?* Agree that it is a graph which shows something happening over a period of time.
- Ask the learners to work with a partner and think of things that happen over time, for example a plant growing, ice melting.
- Sketch the following line graph on the board. Tell the learners that it shows what happens to the level of water at bath time. Discuss what the two axes are showing.
- Point to unmarked positions along the axes and ask the learners to estimate what will go there.

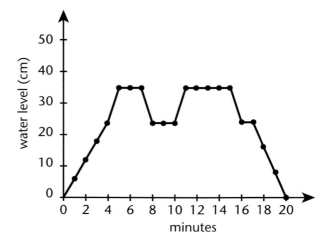

- Ask the learners to talk in pairs about what is happening in the graph. Take feedback.
- Ask: *Look at the times where there is no movement of water. How long do these periods last?*
- Ask: *What is the mode of this data?* Establish that it is the most frequently occurring value. Agree that the mode is 35 cm.
- Write sets of numbers on the board, repeating some in each set. Ask: *What is the mode?*
- Ask the learners to do photocopiable page 107.

Plenary

- Ask the learners to show the line graphs they made in the photocopiable activity. Discuss how the learners found the mode of the data.

Success criteria

Ask the learners:

- What does a line graph show that other graphs don't show?
- Can you give examples of information that can be found on one of these graphs?
- What is meant by the mode?

Ideas for differentiation

Support: Organise these learners to work with a partner of a higher-attaining level.

Extension: Organise these learners to work with a lower-wattaining learner.

Name: _____

Temperature

This table shows temperatures over a 12-hour period:

12:00	13:00	14:00	15:00	16:00	17:00	18:00	19:00	20:00	21:00	22:00	23:00
24°C	25°C	25°C	24°C	24°C	23°C	23°C	22°C	21°C	20°C	17°C	17°C

1.   Work with a partner.

2.   On squared paper draw a line graph to show this data.

3.   What is the mode of the data?

_____

4.   Estimate what the temperature was at:

   19:30 _____

   13:15 _____

   17:45 _____

   22:30 _____

   12:15 _____

   19:45 _____

5.   Now make up some questions to ask the class. Write your questions below.

_____

_____

_____

_____

_____

_____

_____

_____

Probability

- Describe the occurrence of familiar events using the language of chance or likelihood. (5Db1)
- Use ordered lists and tables to help solve problems systematically. (5Ps5)
- Explain methods and justify reasoning orally and in writing; make hypotheses and test them out. (5Ps9)

Set of follow-me cards from photocopiable page 11; 0 to 9 digit cards made from photocopiable page 9; photocopiable page 109; dice.

Starter

- Tell the learners they will be rehearsing multiplication and division facts.
- Share out the follow-me cards between small groups. Keep one card. Read out the multiplication, for example: *4 × 6*. The group with the answer calls it out, then reads the multiplication on their card. Continue.
- Start with the answer on your card. The group that has the multiplication question that goes with it reads it out, then reads out the next answer and so on.

Main activities

- Ask: *What is meant by the word 'probability'?* Establish that this is to do with how likely it is that something will happen. Introduce the vocabulary 'certain', 'likely', 'unlikely' and 'impossible'.
- Ask the learners to discuss in pairs various events certain to happen, for example that afternoon will follow morning. Take feedback.
- Repeat this for events that are likely, unlikely and impossible.
- Choose one of each set of events and place it on a probability line as below:

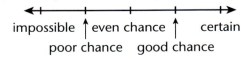

impossible ↑ even chance ↑ certain
    poor chance   good chance

- Invite the learners to think of other events that could be positioned in between those written.
- Show these digit cards: 2, 2, 2, 3, 3 and 3. Ask the learners to say whether there is an even or no chance of the following being picked: 6, 2, 3. Pick a card, shuffle the cards and pick again. Repeat, making a note of the cards selected. Discuss why there was an even chance of picking 2 and 3, why there was no chance of picking 6, and why the chance of picking 2 (or 3) was not certain.
- Ask the learners to work through photocopiable page 109.

Plenary

- Take feedback from the photocopiable activity. Discuss the outcomes by asking why it is likely or unlikely that certain totals will be thrown.
- Discuss when probability is used in real life, for example weather forecasts, football wins / defeats.
- Discuss the probability scale and the idea of even chance.

Ask the learners:

- If we throw two dice, why are 2 and 12 unlikely outcomes? Why is 7 a likely outcome?
- How would you explain probability to someone who didn't know? Is there another way?
- What words are associated with probability? Are there any others?

Support: During the photocopiable activity, reduce the number of throws to eight each. Work with these learners to discover the likelihood of different totals.

Extension: During the photocopiable activity, challenge these learners to explore the probabilities of throwing certain totals with three dice.

Name: _____

Probability

You will need:

Two dice and a partner.

What to do

- The table below lists all the totals that can be thrown using two dice.

- Take it in turns to throw two dice and find the total of the numbers thrown. Record them in the table.

- Keep a tally to show how many times you have thrown each total.

- Do this 15 times each.

Possible totals	Dice thrown	Tally to show number of times the total has been thrown
2		
3		
4		
5		
6		
7		
8		
9		
10		
11		
12		

- From your results complete these statements:

  a) It is likely that I will throw _____

  b) This is a likely total because _____

  c) It is unlikely that I will throw _____

  d) This is an unlikely total because _____

- Write down two totals it will be impossible to throw: _____

- Why is this? _____

- Write down why you cannot be certain of throwing a particular total:

_____

Unit assessment

- What is a frequency table?
- Describe how you would make a pictogram.
- What is the difference between a bar line chart and a pictogram?
- How is a line graph different from a bar line chart?
- What can you tell me about probability?

Summative assessment activities

Observe the learners while they take part in these activities. You will quickly be able to identify those who appear to be confident and those who may need additional support.

Frequency tables

This activity assesses the learners' understanding of frequency tables.

You will need:

Paper; pencils; rulers; two dice for each group.

What to do

- Organise the learners into groups of four.
- Together think of eight popular foods.
- Ask the learners to draw a table with two columns and list these foods in the left-hand column.
- The learners take it in turns to throw the dice.
- They total the two numbers thrown. This number is then allocated to the first food. They put this in the right-hand column.
- They continue to generate numbers until their table is complete.
- Ask them to think of suitable headings for each column and to add these.
- They then give you information from their table, for example *x* is the most popular food, there were *x* more votes for *y* than *z*.

Pictograms

This activity assesses the learners' understanding of pictograms.

You will need:

Paper; pencils; rulers; three dice for each group.

What to do

- Organise the learners into groups of three or four.
- Together think of eight sports. Write these in a frequency table.
- The learners take it in turns to throw the dice.
- They total the three numbers thrown. This number is then allocated to the first sport. Add this to the frequency table.
- Continue to do this until the frequency table is complete.
- They then draw a pictogram to show the information.
- Ask them questions from their pictograms, for example which is the most popular sport, how many more votes for *x* than *y*?

Distribute copies of photocopiable page 111. Ask the learners to read the questions and write the answers. They should work independently.

Name: _____

Working with data

1. This frequency table shows the number of learners in three classes.

   There are 90 learners altogether.

   Work out the number of learners in Class 2.

Class	Number of learners
1	27
2	
3	32

2. This pictogram shows the favourite sports in a class.

   Write down five pieces of information that it gives.

Swimming	☺☺☺☺☺☺☺☺☺
Football	☺☺☺☺☺☺☺☺☺☺
Basketball	☺☺☺☺
Rugby	☺☺☺☺☺☺
Running	☺☺
Gymnastics	☺☺☺☺

☺ = 2 people

1. _____

2. _____

3. _____

4. _____

5. _____

3. Here is a line graph. It shows the temperatures over a 12-hour period.

   Write down five pieces of information that it gives.

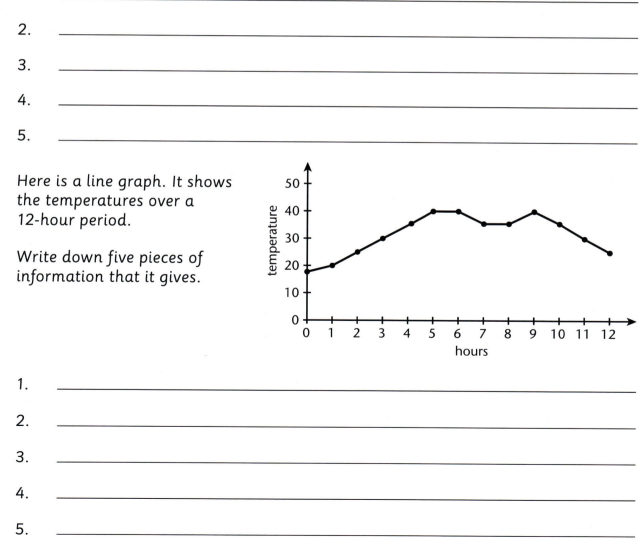

1. _____

2. _____

3. _____

4. _____

5. _____

Unit 2C: Measures and problem solving

Measures

Learning objectives

- Convert larger to smaller metric units (decimals to one place), e.g. change 2.6 kg to 2600 g. (5MI2)
- Round measurements to the nearest whole unit. (5MI4)

Resources

Pencils; paper; photocopiable page 113; scissors.

Starter

- Explain that the learners will practise rounding to help them in the plenary part of the lesson.
- Write four digits on the board. Ask the learners to write down as many four-digit numbers as they can make from those on the board in one minute.
- Ask the learners to round the numbers they have made to the nearest 10, 100 and 1000.

Main activities

- Ask: *What do we measure? Discuss this with your partner.* Agree such things as length, mass, capacity, time, temperature, area.
- Ask: *What units do we use to measure length? What about mass and capacity?* Write the words they say on the board under the headings 'length', 'mass' and 'capacity'. Encourage them to think of imperial and metric units if appropriate.
- For each unit listed, discuss the abbreviation that can be used instead of writing the whole word, for example 'km' for 'kilometre'.
- Ask the learners for equivalences for each category of measure, for example 10 mm = 1 cm, 100 cm = 1 m, 1000 m = 1 km. Ask questions such as: *How many centimetres in 120 mm, 60 mm, 300 m?*

- Write 3.4 kg on the board. Ask the learners what this means. Establish that it is three whole kilograms and point four of a kilogram. Discuss the idea of the 4 representing 400 thousandths because there are 1000 grams in a kilogram.
- Practise writing units for each area in whole and decimal amounts, for example 1.2 m = 1 m 20 cm = 120 cm, 1.6 litres = 1 litre 600 ml = 1600 ml.
- Ask the learners to play the game on photocopiable page 113.

Plenary

- Discuss the equivalences made during the photocopiable activity. Write some of these on the board.
- Take examples of the units and ask the learners to round them to the nearest 10, 100 and 1000.

Success criteria

Ask the learners:

- What units do we use to measure length? Mass? Capacity? Are there any others?
- How many grams are there in 6 kilograms? How do you know?
- How else could we write 5 litres 125 ml? Is there another way?
- What is 256 cm rounded to the nearest 10? How do you know?

Ideas for differentiation

Support: During the photocopiable activity, ask these learners to match two cards of equivalent measures, focusing on mixed measures and whole smaller units. Once they have done this, work with them to add the decimal amount.

Extension: When these learners finish the photocopiable activity, challenge them to make their own sets of units.

Name: _____

Measures

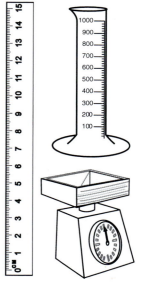

1. Cut out the units cards at the bottom of the page.
   Sort them into groups for length, mass or capacity.

2. Now put the amounts that measure the same together.

3. Complete this table. Put equivalent measures in the same row.
   An example has been done for you.

1 kg 750 g	1.75 kg	1750 g

✄

2 kg 550 g	1 km 500 m	2550 g	10050 ml	1 m 10 cm
25 mm	6 kg 75 g	12 litres 450 ml	12 m 45 cm	103 mm
2 cm 5 mm	2.65 litres	2 litres 650 ml	2.55 kg	10 litres 50 ml
6075 g	2 litres 650 ml	12 450 ml	1500 m	1.1 m
12.45 litres	110 cm	2.5 cm	6.075 kg	10.3 cm
1245 cm	10 cm 3 mm	1.5 km	10.05 litres	12.45 m

Length

Learning objectives

- Read, choose, use and record standard units to estimate and measure length, mass and capacity to a suitable degree of accuracy. (5Ml1)
- Draw and measure lines to the nearest centimetre and millimetre. (5Ml7)
- Consider whether an answer is reasonable in the context of a problem. (5Pt7)

Resources

Pencils; paper for recording; metre stick; rulers; A4 paper; photocopiable page 115.

Starter

- Explain that the learners will practise finding equivalent units of measure.
- Write several three-, four- and five-digit numbers on the board. Link to units of measure, for example 145 cm, 2546 g, 4156 ml, 246 m. Ask the learners to write these amounts in different ways, for example 145 cm = 1 m 45 cm = 1.45 m.
- Repeat for other numbers.

Main activities

- Recap the vocabulary of length, for example 'long', 'wide', 'centimetre'. Ask: *What would we measure in millimetres?* Repeat this question for centimetres, metres and kilometres. Ask: *Can you think of any other units of length?* They may say, for example, miles, feet, inches. Inform the learners that the two forms of measurement are known as metric and imperial. Compare these two types of unit, for example miles and kilometres (0.6 miles is approximately 1 km).
- Ask the learners to think of the different equipment that can be used to measure length, for example trundle wheel, tape measure, milometer.
- Ask the learners to look at and visualise a metre stick. They estimate the width of their table / desk, writing this down. They measure it to check and write the answer in at least two different ways, for example 90 cm, 0.9 m.

- Repeat this for the thickness of a book, recording in millimetres and centimetres.
- Ask the learners choose three other things in the classroom to estimate, measure and record in this way.
- Now ask them to look at the five measurements and order them from shortest to longest. They should draw lines on their paper to represent the lengths that are less than 30 cm.
- Ask the learners to work through photocopiable page 115.

Plenary

- Take feedback from the photocopiable activity. Invite pairs to show their routes and explain how they worked out the distances.
- Finish the lesson by asking the learners to order these lengths from shortest to longest: 1.5 km; 50 cm; 149 metres; 1495 cm; 1842 millimetres; 189 metres.

Success criteria

Ask the learners:

- Which is the longer length, 1842 millimetres or 189 centimetres? How do you know?
- What imperial measurements do you know? Which metric measurements do you know?
- What would you give as an estimate for the width of this book? Why do you think that?

Ideas for differentiation

Support: During the photocopiable activity, ask these learners to measure to the nearest centimetre.

Extension: During the photocopiable activity, ask these learners to measure to the nearest millimetre. Once they have worked out three routes, ask them to find a map on the internet or in an atlas and work out distances from place to place.

Length

1. Work with a partner. Choose two places, one on the west side of the map and one on the east.

   Find a route from one to the other. Work out the distance between the two.

2. Now find two other ways to get from one to the other – a longer distance and a shorter one.

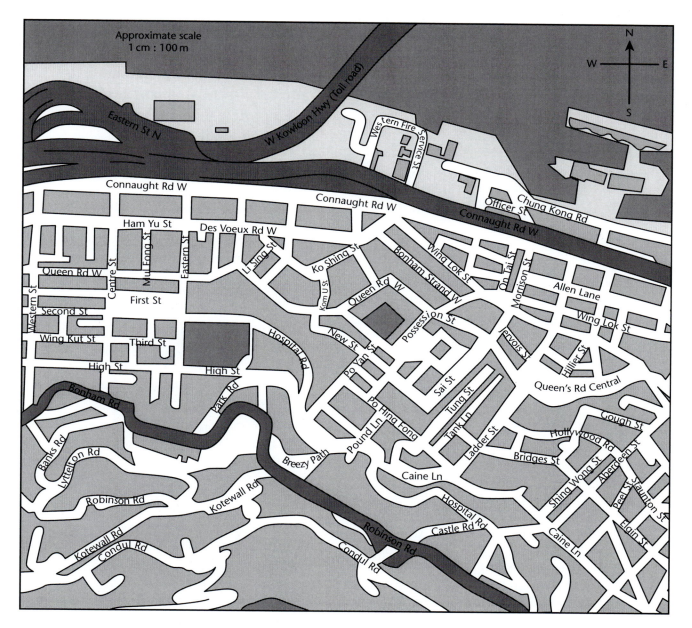

Redrawn after www.chinatouristmaps.com/travel/hong-kong/street-map/hong-kong-island.html

Mass

- Read, choose, use and record standard units to estimate and measure length, mass and capacity to a suitable degree of accuracy. (5MI1)
- Convert larger to smaller metric units (decimals to one place), e.g. change 2.6 kg to 2600 g. (5MI2)
- Interpret a reading that lies between two unnumbered divisions on a scale. (5MI5)

Pencils; paper; a book; 1 kg bag of rice (or similar) for each group of four; weighing scales for each group of four; photocopiable page 117.

Starter

- Explain that the learners will rehearse place value of five- and six-digit numbers.
- Choose different starting numbers linking to units of measure, for example 12 550 ml. Ask: *What does the 1 represent? What about the first 5? How else can we write 12 550 ml?*
- Repeat.

Main activities

- Recap the vocabulary of mass, for example 'heavy', 'light', 'gram'. Ask: *What would we measure in grams?* Repeat for kilograms. *Can you think of any other units of mass?* They may say, for example, stones, pounds. Ask: *What are these two forms of measurement known as?* Agree metric and imperial.
- Ask the learners to think of the different equipment that can be used to measure mass, for example kitchen scales, bathroom scales, pan balance.
- Ask them to look at a book and estimate what it weighs. Discuss the difficulty of doing this if they can't feel it and haven't got a known mass to compare with.

- Invite eight learners to feel the kilogram bag of rice. They then hold the book and estimate its mass. Write their estimates on the board. Ask them to weigh it to check. As a class, decide who gave the closest estimate.
- Give groups of four learners a bag of rice and a set of scales. Ask them to choose two items from the classroom. They each hold the rice to get a feel for 1 kg, then hold the items and make an estimate. They weigh their items and compare the results with their estimates.
- Ask the learners to work through photocopiable page 117.

Plenary

- Invite groups to share what they did in the photocopiable activity. They should tell the class which items they weighed, the estimates they gave and the actual mass.
- Ask the learners to assess their confidence in estimating and measuring mass.

Ask the learners:

- What units do we use to measure mass? Are there others?
- What would we measure in kilograms? What else?
- How many grams are there in 5.2 kg? How do you know?
- How can you make a sensible estimate of the mass of an object? Is there another way?

Support: Ensure these learners are working in a group of mixed ability so they can be supported by their peers.

Extension: Encourage these learners to support the others in their groups.

Name: _____

Mass

You will need:

A 1 kg bag of rice and weighing scales.

What to do

● Work in a group of four.

● Find five objects from around the classroom.

● Take one object at a time. Use the bag of rice to help you estimate its mass.

● Together agree on your estimate and write it in the table below.

● Next weigh your object and then work out the difference between your estimate and the actual mass. How close were you?

● Complete the table. Do this for all the objects.

Object	Estimate	Actual	Difference

● Show the mass of each object on these scales:

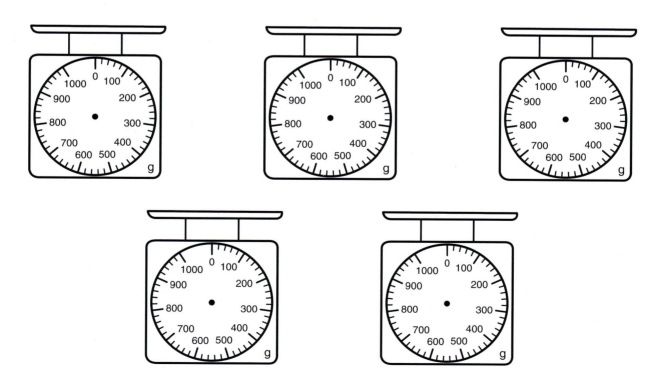

Capacity

- Read, choose, use and record standard units to estimate and measure length, mass and capacity to a suitable degree of accuracy. (5MI1)
- Convert larger to smaller metric units (decimals to one place), e.g. change 2.6 kg to 2600 g. (5MI2)
- Order measurements in mixed units. (5MI3)
- Compare readings on different scales. (5MI6)

Pencils; paper; glass; water; different measuring jugs and cylinders; different containers, including one 1 or 2 litre bottle for each group of four; photocopiable page 119.

Starter

- Explain that the learners will order amounts from smallest to largest.
- Write a selection of mixed units of length on the board, for example 1245 cm, 124 mm, 12.4 m, 1 km 245 m. Ask the learners to write them in order. Ask: *How do you know you have the correct order?*
- Repeat for mass and capacity.

Main activities

- Recap the vocabulary of capacity, for example 'full', 'empty', 'litre'. Discuss what capacity is (the amount a container can hold). Compare this with volume (the amount in a container). Ask: *What would we measure in millilitres?* Repeat for litres. Ask: *Can you think of any other units of capacity?* They may say, for example, gallons, pints. Ask: *What are these two forms of measurement known as?* Agree metric and imperial.
- Ask the learners to think of the different equipment that can be used to measure capacity, for example measuring jug or cylinder.
- Show a glass and ask the learners to estimate its capacity. Discuss the difficulty of doing this if they haven't got a known capacity to compare with it.

- Show a litre bottle of water. Ask the learners to compare the two and then estimate the capacity of the glass. They should write their estimates down. Pour the water from the bottle into the glass. Then measure the amount into different measuring jugs and cylinders. Compare the different scales. Agree that the amount looks different but is the same.
- Compare the learners' estimates with the actual amount.
- Repeat this with different containers in the classroom.
- Ask the learners to work through photocopiable page 119.

Plenary

- Invite groups to share what they did in the photocopiable activity. They should tell the class which containers they used, the estimates they gave and the actual capacities.
- Ask the learners to assess their confidence in estimating and measuring capacity.

Ask the learners:

- What units do we use to measure capacity? Are there others?
- What would we measure in litres? What else?
- How many millilitres are there in 2.1 litres? How do you know?
- How can you make a sensible estimate of a container's capacity? Is there another way?

Support: For the photocopiable activity, ensure these learners are working in a group of mixed attainment so they can be supported by their peers.

Extension: During the photocopiable activity, encourage these learners to support the others in their groups.

Capacity

You will need:

A 1-litre or 2-litre bottle of water and a measuring jug.

What to do

- Work in a group of four.

- Find four small containers in your classroom.

- Take one container at a time. Use the bottle of water to help you estimate its capacity.

- Together agree on your estimate and write it in the table below.

- Next fill your container with water, then pour that water into a measuring jug and find the capacity.

- Work out the difference between your estimate and the actual capacity. How close were you?

- Complete the table for all the containers.

Object	Estimate	Actual	Difference

- Show the four capacities on these measuring cylinders:

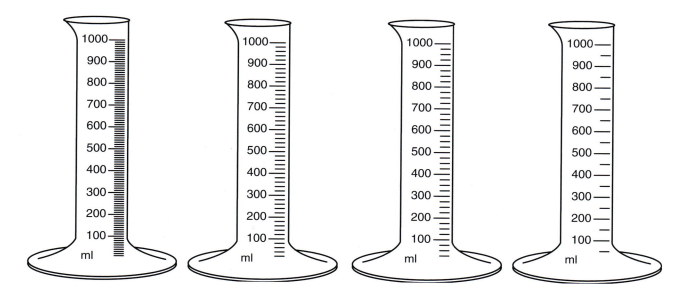

Time 1

- Recognise and use the units for time (seconds, minutes, hours, days, months and years). (5Mt1)
- Use a calendar to calculate time intervals in days and weeks (using knowledge of days in calendar months). (5Mt5)
- Understand everyday systems or measurement in length, weight, capacity, temperature and time and use these to perform simple calculations. (5Pt1)

Pencils; paper; photocopiable page 192; set of cards from photocopiable page 121 for each group of four (some cards are blank so you can add your own words).

Starter

- Explain that the learners will be answering questions about units of time.
- Give the learners a copy of photocopiable page 192. Ask questions from it that involve, for example, putting their finger on 2nd February and counting on 7 days, 70 days, 24 hours and so on. Encourage them to work out the most efficient way to do this, for example counting 10 weeks instead of 70 days.
- Ask questions such as: *How long is it from 28th March to 19th August?*

Main activities

- Ask: *What words can you tell me to do with time?* Give the learners a minute to talk in pairs, then take feedback. Write the vocabulary they say on the board. Ensure these are included: year, month, week, day, hour, minute, second, analogue, digital, 24-hour time. Discuss what each word means.
- Give the learners a word from the list and ask them to make up a sentence to show what it means to share with the class.

- Divide the class into groups of four. Number the learners in each group from 1 to 4: number 1s should be the most confident at explaining, number 4s the least confident. Give each group a set of 'just a minute' cards from photocopiable page 121. Each learner has a minute to explain the words on as many of the cards as they can. Begin with the number 1s. When the others in the group guess correctly the card is turned face up on the table. After a minute, they count how many they guessed. These are put together and placed on the top of the pile. The number 2s then take their turn, beginning with those already guessed correctly. Repeat this until everyone has had a turn at explaining.

Plenary

- Play the just a minute game as a class: you give the meanings, the learners call out the answers. Do this three times. Are they faster by the end of the third turn?
- Recap the words they have been working with and their meanings.

Ask the learners:

- What is an analogue clock? How is it different from a digital one?
- If my watch says 25 to 2, what would the time be on a digital one? How do you know?
- How many hours are there in one day? How many in five days? How did you work that out?

Support: Make these learners number 4s during the just a minute game. This will enable them to listen to the explanations of the words three times before explaining them themselves.

Extension: Make these learners number 1s during the just a minute game. They need to devise clear and precise definitions.

Just a minute cards

millennium	month	minute
century	week	second
decade	day	analogue clock
year	hour	digital clock
yesterday	today	tomorrow
12-hour time	24-hour time	soon

Time 2

- Tell and compare the time using digital and analogue clocks using the 24-hour clock. (5Mt2)
- Calculate time intervals in months or years. (5Mt6)
- Solve a larger problem by breaking it down into sub-problems or represent it using diagrams. (5Ps10)

Resources

Photocopiable page 192; pencils; paper; class clock; individual analogue clock face for each learner; photocopiable page 123.

Starter

- Tell the learners that they will rehearse finding time intervals.
- Give each pair of learners a copy of photocopiable page 192. Ask them to work out different time intervals, for example from 12th December to 30th April.
- Give problems such as: *The plans for a new building take four weeks to design, the building takes 18 months, the decorating two weeks. If we start on 1st February next year, when will it be finished?*

Main activities

- Give each learner, or pair of learners, an analogue clock face. Ask them to show you different times and write down the equivalent digital time. For example, say: *20 minutes to six in the evening.* The learners should show that time and write 5:40p.m. and, if appropriate, 17:40.
- Set problems such as: *Mandy left home at 10:30a.m. She arrived at the shopping centre 40 minutes later. Show me on your clocks the time she got to the shopping centre.*
- Tell the learners that a time number line is helpful to work out time differences. It is also helpful when converting 12-hour clock time to 24-hour clock time. Draw this on the board:

- Give some 12-hour clock times and ask the learners to plot them on the number line and then use it to convert to 24-hour clock times, for example 7:15a.m., 9:40p.m.
- Set more problems and demonstrate how to use the number line to solve them, for example: *Awatif began her homework at 15:10. She finished at 16:55. How long did it take?*

15:10 to 16:00 is 50 mins, 16:00 to 16:55 is 55 mins.

Total = 105 mins or 1 hour 45 minutes

- Ask the learners to work through the problems on photocopiable page 123.

Plenary

- Take feedback from the photocopiable activity. Invite the learners of different attainment levels to share how they worked out the answers to the problem.
- Ask the learners to share any problems they made up for the rest of the class to solve.

Success criteria

Ask the learners:

- What is 5:15p.m. as a 24-hour clock time? How did you work that out? Is there another way?
- What is the time difference between 09:40 and 13:25? How did you work that out? Is there another way?
- Why is a number line helpful for working out time differences? Can you explain in a different way?

Ideas for differentiation

Support: Adapt the times on photocopiable page 123 so that these learners are working out differences between whole, half and quarter hour intervals.

Extension: Adapt the times on photocopiable page 123 so that these learners are working out differences between minute intervals.

12:00 1:00 2:00 3:00 4:00 5:00 6:00 7:00 8:00 9:00 10:00 11:00 12:00 1:00 2:00 3:00 4:00 5:00 6:00 7:00 8:00 9:00 10:00 11:00
00:00 01:00 02:00 03:00 04:00 05:00 06:00 07:00 08:00 09:00 10:00 11:00 12:00 13:00 14:00 15:00 16:00 17:00 18:00 19:00 20:00 21:00 22:00 23:00

Time problems

1. Solve these problems. Use the number lines to help you.

   a) Abdul went for a walk.

   He left his house at 14:20 and was out for 2 hours 25 minutes.

   When did he get home?

   | 14:00 | 14:30 | 15:00 | 15:30 | 16:00 | 16:30 | 17:00 |

   My answer: _____

   b) The Formula 1 car race started at 14:10 and finished at 16:25.

   How long did the race last?

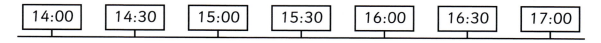

   | 14:00 | 14:30 | 15:00 | 15:30 | 16:00 | 16:30 | 17:00 |

   My answer: _____

   c) Dan and Wally watched their favourite football team play a match on the television.

   The programme started at 15:05 and finished at 17:40.

   How long did the programme last?

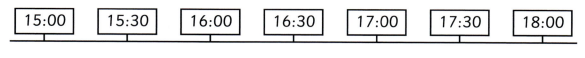

   | 15:00 | 15:30 | 16:00 | 16:30 | 17:00 | 17:30 | 18:00 |

   My answer: _____

   d) Fiona and Seema began their homework at 15:10.

   Fiona finished her homework after an hour. Seema finished hers at 16:55.

   How much longer did it take Seema to do her homework?

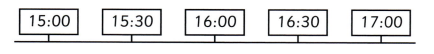

   | 15:00 | 15:30 | 16:00 | 16:30 | 17:00 |

   My answer: _____

2. Now make up and solve some problems of your own. Use the back of this paper.

Time 3

Learning objectives

- Read timetables using the 24-hour clock. (5Mt3)
- Calculate time intervals in seconds, minutes and hours using digital or analogue formats. (5Mt4)
- Explain methods and justify reasoning orally and in writing; make hypotheses and test them out. (5Ps9)

Resources

Pencils; paper; individual analogue clock face for each learner; photocopiable page 125.

Starter

- Tell the learners they will practise partitioning and counting on to find totals and differences to help in the lesson.
- Write two three-digit numbers on the board, for example 245 and 178. The learners find the total by partitioning the second number and adding each part onto the first (245 + 100 + 70 + 8). They find the difference by counting on (178 + 22 + 45, 22 + 45 = 67).
- Repeat this with several pairs of numbers.

Main activities

- Draw this number line, as in Time 2, on the board:

```
 12:00 1:00  2:00  3:00  4:00  5:00  6:00  7:00  8:00  9:00 10:00 11:00
◄──┼────┼────┼────┼────┼────┼────┼────┼────┼────┼────┼────┼──
 00:00 01:00 02:00 03:00 04:00 05:00 06:00 07:00 08:00 09:00 10:00 11:00

 12:00 1:00  2:00  3:00  4:00  5:00  6:00  7:00  8:00  9:00 10:00 11:00
───┼────┼────┼────┼────┼────┼────┼────┼────┼────┼────┼────┼──►
 12:00 13:00 14:00 15:00 16:00 17:00 18:00 19:00 20:00 21:00 22:00 23:00
```

- Give some 12-hour clock times and ask the learners to plot them on the number line and then use it to convert to 24-hour clock times, for example 7:15a.m., 9:40p.m.
- Ask: *When do we usually see this type of time displayed?* Agree on such things as ovens, mobiles phones and also timetables. Recap what a timetable is: it gives information about, for example, when buses / trains / planes arrive and depart.

- Give the learners an analogue clock face each. Ask them to show you different times. They should write down the equivalent 24-hour clock time for each. For example, say: *15 minutes past 8 in the evening.* The learners show that time and write down 20:15.
- Set problems such as: *The plane from London to Cairo takes 5 hours 36 minutes. It takes off at 14:17. When does it land?* The learners should draw a time number line, beginning at 14:17. They use this to count on 5 hours and then 36 minutes to find the answer (19:53 or 7:53p.m.).
- Ask the learners to work through the problem on photocopiable page 125.
- Answers to photocopiable activity:
  7 h 15 min, 6 h 25 min, 12 h 20 min,
  8 h 5 min, 8 h 50 min, 7 h, 22 h 20 min,
  11 h 25 min.

Plenary

- Take feedback from the photocopiable activity, inviting the learners to share their strategies. Discuss why some flights took longer than others, for example stopovers in other countries.
- Ask the learners to assess their confidence in finding totals and differences of 24-hour clock times.

Success criteria

Ask the learners:

- What is 18:24 as a 12-hour clock time? How did you work that out? Is there another way?
- What is the time difference between 12:36 and 17:05? How did you work that out? Is there another way?
- What flight should Faisal's father catch? Why do you think that?

Ideas for differentiation

Support: Adapt the times on photocopiable page 125 so that these learners are working out differences between whole, half and quarter hour intervals.

Extension: Adapt the times on photocopiable page 125 so that these learners are working out differences between minute intervals.

Name: _____

Timetable

The table below shows the departure times from
Oman International Airport and the arrival times at
London Heathrow.

1.   How long does each flight take?

2.   Draw number lines on paper to help you.

Flight	Depart Oman International	Arrive London Heathrow	Journey time
1	00:25	07:40	
2	00:55	07:20	
3	00:55	13:15	
4	05:10	13:15	
5	09:50	18:40	
6	11:15	18:15	
7	12:25	10:45	
8	19:55	07:20	

3.   Faisal's father needs to get to a meeting at 16:00 in London. He is not keen on
flying so he would like a short flight. Which one should he choose?

_____

4.   Why do you think some of these flights take so long and others are much shorter?

_____

_____

Perimeter and area 1

Learning objectives

- Measure and calculate the perimeter of regular and irregular polygons. (5Ma1)
- Choose an appropriate strategy for a calculation and explain how they worked out the answer. (5Ps2)

Resources

Set of follow-me cards from photocopiable page 11; pencils; centimetre squared paper; rulers; photocopiable page 127.

Starter

- Tell the learners that they will rehearse multiplication and division facts.
- Share out the follow-me cards between small groups. Keep one card. Read out the multiplication, for example: *4 × 6*. The group with the answer calls it out, then reads the multiplication on their card. Continue.
- Repeat but start with the answer on your card so that they are considering division facts.

Main activities

- Ask: *What do we mean by perimeter and area? Talk to your partner.* Take feedback. Agree that area is the amount of space something takes up, perimeter is the outside of that area.
- Ask the learners to give examples, for example a classroom (area: floorspace, perimeter: walls).
- Recap that the perimeter is a length and so will be measured in millimetres, centimetres, metres and kilometres. Area is measured in squares of these units.
- Discuss what areas might be measured in square millimetres (stamp), centimetres (book), metres (football pitch) and kilometres (country).

- Ask the learners to draw a triangle on squared paper. Ask: *How can you find the perimeter and area of the triangle?* Agree that they measure the sides with a ruler and total them for the perimeter. Agree that they can count the squares, matching up part squares that are approximately a whole, for the area. Ask the learners to do this. Take feedback, ensuring the learners use the correct units.
- Ask them to draw other triangles and irregular shapes on their paper and find their perimeters and areas.
- Ask the learners to complete photocopiable page 127.

Plenary

- Take feedback from the photocopiable activity. Ask individuals to explain how they found the perimeters and areas of the two shapes. Did they find a quick way to count the squares?

Success criteria

Ask the learners:

- What can you tell me about perimeter? What else?
- What can you tell me about area? What else?
- Where do we see areas in real life? What about perimeters?
- If I draw a triangle which has sides of 12 cm, what is its perimeter? How did you work that out?

Ideas for differentiation

Support: During the photocopiable activity, allow these learners to concentrate on the first shape.

Extension: When they have completed the photocopiable activity, challenge these learners to make up their own shape and estimate its perimeter and area.

Perimeter and area

Estimate the perimeter and area of these shapes. Each square represents 1 square centimetre.

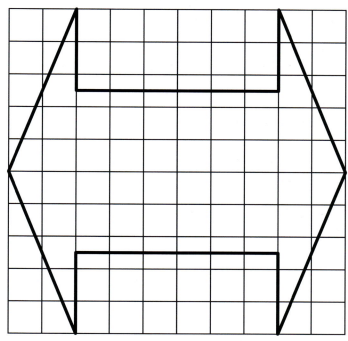

My estimate of the perimeter is _____

My estimate of the area is _____

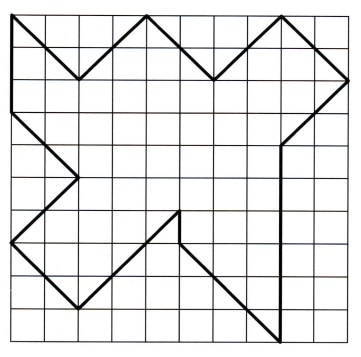

My estimate of the perimeter is _____

My estimate of the area is _____

Perimeter and area 2

Learning objectives

- Measure and calculate the perimeter of regular and irregular polygons. (5Ma1)
- Use the formula for the area of a rectangle to calculate the rectangle's area. (5Ma3)
- Deduce new information from existing information to solve problems. (5Ps4)

Resources

Pencils; paper for recording; rulers; centimetre squared paper; photocopiable page 129.

Starter

- Explain to the learners that they will be rehearsing mental addition and subtraction strategies.
- Call out two three-digit numbers. They find their total and difference, making jottings if necessary. Invite volunteers to share their strategies, such as partitioning and complementary addition.
- Repeat.

Main activities

- Recap what is meant by area and perimeter and how they are measured. Ask: *Who can remember the quick way to find the perimeter of a rectangle?* Invite volunteers to explain. Agree: add the length and width then double or double each side and then add. Ask: *What could we do if we had a square?* Agree that you could multiply one length by 4.
- Ask the learners to draw some rectangles on a piece of squared paper. They choose the lengths of the shapes' sides. Ask them to work out their perimeters using one of the methods discussed.
- Focus on area. Ask: *How can you find the area of your rectangles without counting the squares?* Demonstrate with this rectangle:

The rectangle is 4 cm long and 2 cm wide. Four lots of two is 8, so the area is 8 square centimetres.

Agree that the area can be found using the formula *l* × *w* (length × width).

- Now ask the learners to find the areas of their rectangles.
- Ask the learners to draw different rectangles with a perimeter of 24 cm. Once they have drawn at least four, they should work out their areas. Ask: *What does this tell us about perimeter and area?* Agree that the same perimeter can surround different areas.
- Ask the learners to complete photocopiable page 129.

Plenary

- Discuss the work from the photocopiable activity. Invite individuals to share the lengths and widths of their shapes, and ask the class to work out their areas.
- Ask the learners to assess their confidence at finding perimeters and areas of rectangles using the formula.

Success criteria

Ask the learners:

- How can you find the area of a rectangle? Is there another way? What about the perimeter?
- If I draw a rectangle which has sides of 15 cm and 20 cm, what is its perimeter? How did you work that out? What is its area? How did you work that out?

Ideas for differentiation

Support: Allow these learners to draw their shapes on squared paper. They can add side lengths to find perimeters and count squares to find areas.

Extension: Challenge these learners to explore the different perimeters and areas they can make using whole and half centimetres.

Name: _____

Area and perimeter

You will need:

A partner, a pencil and a ruler.

What to do

● Draw two squares and two rectangles in the space below.

● Measure their lengths and widths in whole centimetres. Make sure you are accurate when you use your ruler.

● Make a note of the lengths and widths on a separate piece of paper. Don't label your shapes!

● Now give your shapes to your partner. Ask them to work out the area of your shapes using the formula $l \times w$ (length × width).

● When your partner has done this, take your shapes back. Check their answers to see if you agree.

Unit assessment

Questions to ask

- Tell me some equivalent lengths (for example 1.4 km = 1 km 400 m = 1400 m).
- Which units do we use to measure capacity?
- What measuring equipment can we use to measure mass?
- Is it true that as the perimeter gets longer the area gets larger?
- What is meant by area?

Summative assessment activities

Observe the learners while they take part in these activities. You will quickly be able to identify those who appear to be confident and those who may need additional support.

What's the time?

This activity assesses the learners' understanding of analogue and digital time.

You will need:

Paper; pencils; individual clock faces.

What to do

- Organise the learners into groups of four.
- Give each learner a clock face.
- Call out various times, for example: *20 minutes to 5*.
- The learners should show these times on their clock faces and write them digitally on paper.
- Encourage them to write 24-hour clock times if you think they can.

Timetables

This game assesses the learners' knowledge of finding time differences.

You will need

Pencils; paper; copy of timetable below for each learner.

Flight	Depart Quito International, Ecuador	Arrive Schiphol International, Amsterdam
	All times are Ecuadorian	
1	13:35	03:40
2	15:55	07:20
3	21:10	09:28
4	23:15	11:54

What to do:

- Organise the learners into groups of four.
- Give each learner a copy of the timetable.
- Ask them to work out the flight time for each flight.
- Encourage them to use a time number line.

Written assessment

Distribute copies of photocopiable page 131. Ask the learners to read the questions and write the answers. They should work independently.

Name: _____

Working with measures 2

1. Write 13 kg 450 g in two different ways.

   _____

2. A jug has a capacity of 2.75 litres. A bucket has a capacity of 5 litres. How much more does the bucket hold?

   _____

3. A ribbon is 1.6 m long. Some string is 1 m 6 cm in length.

   a) Which is the longest?

   _____

   b) How do you know?

   _____

4. Draw a clock face to show 23:10.

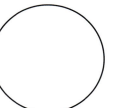

5. The film started at 17:25. It finished at 19:40. How long did it last?

   _____

6. Use a time number line to work out the difference between 09:05 and 13:55.

   _____

7. I draw a regular pentagon. Its perimeter is 60 cm. What is the length of its sides?

   _____

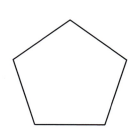

8. A rectangle has sides of 10 cm and 9 cm.

   a) What is its perimeter?

   _____

   b) What is its area?

   _____

Fractions 1

- Order and compare negative and positive numbers on a number line and temperature scale. (5Nn9)
- Calculate a rise or fall in temperature. (5Nn10)
- Relate finding fractions to division and use to find simple fractions of quantities. (5Nn18)
- Deduce new information from existing information to solve problems. (5Ps4)

Pencils; paper; photocopiable page 133; dice for each learner.

Starter

- Write a series of positive and negative numbers on the board, for example: *–12, 5, 23, –3, –9, 8.* Ask the learners to write them in order. Ask them to pick two of the numbers and find their difference, then write a number sentence with the symbol > or <.
- Ask questions relating to temperature, for example: *The temperature in the morning was –2 °C, by lunchtime it was 12 °C. How many degrees had it risen?*
- Do this a few times.

Main activities

- Ask: *What is meant by half?* Establish that it is an amount divided into two equal parts. This could be an object, such as an apple, or a number. Repeat for quarter (four equal parts) and eighth (eight parts). Write these on the board: $\frac{1}{2}$, $\frac{1}{4}$ and $\frac{1}{8}$. Ask: *Which part of the fraction shows how many parts to divide into?* Agree this is the bottom number, the denominator.
- Call out different fractions, for example: $\frac{1}{5}$, $\frac{1}{10}$, $\frac{1}{3}$. Ask the learners to write down how many equal parts there will be, for example 5, 10 and 3.
- Call out different fractions of amounts within a context the learners can relate to. They should write down what the amount would be, for example $\frac{1}{3}$ of 24 cakes = 8 cakes.

- Focus on the top number of the fraction, the numerator. Ask: *What do you think the numerator is?* Agree that it is the number of parts needed. Demonstrate with $\frac{4}{5}$ of 25: $\frac{1}{5}$ is 5, so $\frac{4}{5}$ is 4 of lots of 5, 20. Model this with counters or by drawing 25 circles and showing $\frac{4}{5}$. Repeat for other fractions.
- Call out fractions with different numerators, for example: $\frac{3}{8}$ *of 16 books.* Ask the learners to work out the amount.
- Ask the learners to work through photocopiable page 133.

Plenary

- Invite individuals to share the fractions they made. Ask them to tell the class the two-digit numbers they chose and share their strategies for finding the fractions of these numbers.
- Ask the learners to share the problems they made up.

Ask the learners:

- What can you tell me about fractions? What else?
- What are the words used to describe the numbers in a fraction?
- Can you explain how to find $\frac{1}{4}$ of 48? Is there another way?
- If we had to find $\frac{3}{4}$ of 48 what would we do? Can you do this in another way?

Support: During the photocopiable activity, ask these learners to find halves and quarters of multiples of 4 that you give them. Encourage them to use a halving strategy to find the unit fraction.

Extension: During the photocopiable activity, challenge these learners to work out three-digit multiples of the denominator.

Name: _____

Fractions of numbers

$$\frac{1}{2} \quad \frac{3}{4} \quad \frac{1}{8}$$

$$\frac{2}{3} \quad \frac{5}{6} \quad \frac{4}{5}$$

You will need:

A partner and two dice.

What to do

- Take it in turns to throw the dice. Use the two numbers you throw to make a fraction. The numerator needs to be smaller than the denominator.

- Write down a two-digit number that is a multiple of the denominator.

- Now find the fraction of the two-digit number. Record what you do in the table below. An example has been done for you:

Dice thrown	Fraction made	Two-digit number	Amount
6 and 5	$\frac{5}{6}$	54	$\frac{1}{6}$ of 54 is 9 so $\frac{5}{6}$ of 54 is 45

- Choose some of your fractions and make up a problem for each one. Write it on the back of this paper.

Fractions 2

- Recognise equivalence between: $\frac{1}{2}$, $\frac{1}{4}$ and $\frac{1}{8}$; $\frac{1}{3}$ and $\frac{1}{6}$; $\frac{1}{5}$ and $\frac{1}{10}$. (5Nn15)
- Choose an appropriate strategy for a calculation and explain how they worked out the answer. (5Ps2)
- Explore and solve number problems and puzzles, e.g. logic problems. (5Ps3)

Pendulum (three interlocking cubes on a piece of string or similar); A4 paper cut into 3 cm strips lengthways, enough for 11 strips per learner; pencils; paper for recording; photocopiable page 135; rulers.

Starter

- Explain that the learners will practise counting in steps of different sizes.
- Show the pendulum. Swing it from side to side. As you swing it the learners should count in steps of seven from zero to 70 and then back again.
- Repeat for different step sizes, including 0.1 and negative numbers.

Main activities

- Draw five rectangles of the same size on the board. Ask individual learners to use them to explain what $\frac{1}{2}$, $\frac{1}{4}$, $\frac{1}{8}$, $\frac{1}{5}$ and $\frac{1}{10}$ are. They should draw lines on the rectangles to demonstrate:

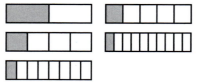

Agree that $\frac{1}{2}$ is one rectangle shared into two parts of equal size, $\frac{1}{4}$ is shared into four, $\frac{1}{8}$ into eight, $\frac{1}{5}$ into five and $\frac{1}{10}$ into ten.

- Discuss which is the smallest fraction and why. Do this by comparing the rectangles that show one-half and one-tenth. Establish that one-tenth is the smallest fraction because there are ten parts. One-half is the largest because there are only two parts.

- Give each student three strips of paper. They fold the first in half, the second in half and half again to make quarters and the third in half, half again and half once more to make eighths. They label each part with the appropriate fraction.
- Ask: *What do we mean by 'equivalent'?* Establish that this is when fractions made using different numbers are exactly the same size. Ask the students to identify equivalent fractions using their strips, for example $\frac{1}{2} = \frac{2}{4} = \frac{4}{8}$, $\frac{1}{4} = \frac{2}{8}$, $\frac{3}{4} = \frac{6}{8} = \frac{1}{2} + \frac{1}{4}$.
- Ask the learners to work through photocopiable page 135.

Plenary

- Take feedback from the photocopiable activity. Ask individuals to share the equivalent fractions they identified.
- Ask the learners to find a pattern in all fraction equivalences for one-half (the denominator is always double the numerator).

Ask the learners:

- What is meant by an equivalent fraction? Can you explain in a different way?
- What fractions are equivalent to one-half? Are there any others?
- What is equivalent to one-quarter? What else?
- Can you see a pattern in the fractions that are equivalent to one-half? What about one-quarter?

Support: During the photocopiable activity, ask these learners to focus on equivalences with one-half.

Extension: When these learners have completed the photocopiable activity, challenge them to explore other fractions such as sevenths, ninths and twelfths.

Fraction walls

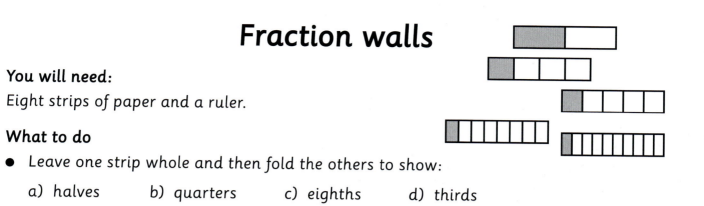

You will need:

Eight strips of paper and a ruler.

What to do

- Leave one strip whole and then fold the others to show:

    a) halves   b) quarters   c) eighths   d) thirds

    e) sixths   f) fifths   g) tenths

- You can make some of these by halving. For others, work out how big each part will be using a ruler.

- Order them with the largest fraction at the top and the smallest on the bottom. You have made a fraction wall!

- Sketch your wall here:

- Now make a list of some of the equivalent fractions that you have made. How many can you find?

    Here is an example: $\dfrac{2}{5} = \dfrac{4}{10}$

_____

_____

_____

Fractions 3

Learning objectives

- Add or subtract near multiples of 10 or 100, e.g. 4387 − 299. (5Nc9)
- Change an improper fraction to a mixed number, e.g. $\frac{7}{4}$ to $1\frac{3}{4}$; order mixed numbers and place between whole numbers on a number line. (5Nn17)
- Investigate a simple general statement by finding examples which do or do not satisfy it, e.g. the sum of three consecutive whole numbers is always a multiple of three. (5Ps8)

Resources

Pencils; paper; photocopiable page 137 for each pair or group; scissors; sets of 0 to 9 digit cards made from photocopiable page 9.

Starter

- Tell the learners they will be revising adding and subtracting near multiples of 10 and 100 to keep this skill fresh in their minds.
- Write some calculations on the board, for example: *156 + 79, 345 − 199*. These need to be answered using the strategy of rounding to the nearest multiple of 10 or 100 and adjusting.
- Invite individuals to share their strategies.

Main activities

- Ask the learners to explain what these fractions are: $\frac{1}{2}, \frac{1}{4}, \frac{1}{8}, \frac{1}{5}$ and $\frac{1}{10}$. Repeat for $\frac{3}{4}, \frac{2}{3}, \frac{5}{8}$. See page 132 for an explanation.
- Recap equivalent fractions, such as $\frac{2}{4} = \frac{1}{2}, \frac{4}{10} = \frac{2}{5}$. Ask the learners to think of some of their own.
- Discuss how many of each makes a whole. Ask the learners to write number sentences to show this, for example $\frac{8}{8} = 1, \frac{5}{5} = 1$.
- Write this fraction on the board: $1\frac{3}{8}$. Ask the learners to tell you how many eighths this is. Agree 11. Ask them to explain how they know. Establish that one whole is $\frac{8}{8}$, so if this is added to $\frac{3}{8}$ the result is $\frac{11}{8}$. Repeat for other mixed numbers with a variety of denominators.

- Give this statement: $\frac{7}{4}$ *is greater than* $\frac{11}{8}$. Ask the learners to prove whether or not this is true by turning them into mixed numbers and then the quarters into eighths. Agree that the statement is true.
- Repeat for other similar fractions.
- Organise the learners into pairs or small groups and ask them to play the game on photocopiable page 137.

Plenary

- Take feedback from the photocopiable activity.
- Write some improper fractions on the board and ask the learners to turn them into mixed numbers. Invite individuals to explain exactly what they have done.
- Repeat for turning mixed numbers into improper fractions.

Success criteria

Ask the learners:

- Can you explain what $2\frac{1}{2}$ is?
- How else can you write $1\frac{3}{5}$? When might you see $1\frac{3}{5}$ in real life?
- Which is larger, $2\frac{3}{8}$ or $2\frac{1}{4}$? How do you know?
- What operations help us to change improper fractions to mixed numbers and vice versa?

Ideas for differentiation

Support: During the game, ask these learners to focus on changing between mixed numbers and improper fractions which involve halves and quarters.

Extension: Challenge these learners to make their own cards for the game, picking two digit cards to make the numerator and a third for the denominator.

Fraction pairs game

What to do

- Play this game with a partner or in a small group.
- Cut out the cards. Shuffle them well and place them face down on the table.
- The first player picks two cards. If they match they keep the pair and have another turn. If they don't match, the player places them both face down on the table.
- The next player takes their turn.
- Play continues until all the cards have been paired.
- The player with the most pairs wins.

✂

$1\frac{1}{2}$	$\frac{3}{2}$	$2\frac{1}{2}$	$\frac{5}{2}$	$3\frac{1}{2}$
$\frac{7}{2}$	$1\frac{1}{4}$	$\frac{5}{4}$	$1\frac{3}{4}$	$\frac{7}{4}$
$2\frac{1}{4}$	$\frac{9}{4}$	$2\frac{3}{4}$	$\frac{11}{4}$	$3\frac{1}{4}$
$\frac{13}{4}$	$3\frac{3}{4}$	$\frac{15}{4}$	$1\frac{1}{8}$	$\frac{9}{8}$
$1\frac{3}{8}$	$\frac{11}{8}$	$2\frac{5}{8}$	$\frac{21}{8}$	$2\frac{7}{8}$
$\frac{23}{8}$	$3\frac{5}{8}$	$\frac{29}{8}$	$1\frac{1}{5}$	$\frac{6}{5}$
$1\frac{3}{5}$	$\frac{8}{5}$	$2\frac{3}{5}$	$\frac{13}{5}$	$2\frac{4}{5}$
$\frac{14}{5}$	$1\frac{7}{10}$	$\frac{17}{10}$	$4\frac{3}{10}$	$\frac{43}{10}$

Decimals

Learning objectives

- Use decimal notation for tenths and hundredths and understand what each digit represents. (5Nn4)
- Round a number with one or two decimal places to the nearest whole number. (5Nn7)
- Order numbers with one or two decimal places and compare using the > and < signs. (5Nn11)

Resources

Counting stick; pencils; paper; photocopiable page 139; set of 0 to 9 digit cards made from photocopiable page 9 for each learner.

Starter

- Tell the learners that they will rehearse counting in steps of different sizes. This may help in the lesson.
- Show the learners the counting stick and explain that one end represents zero. Ask: *What steps will we count in to get from zero to 1?* Agree on 0.1. Together count in these steps from zero to 1 and back.
- Repeat for other decimal steps, for example 0.2.

Main activities

- Ask: *What can you tell me about decimals? Talk to your partner.* Take feedback. Agree that decimals are a type of fraction. They are parts of a whole, for example 0.3. They can also include a whole number, for example 14.5.
- Discuss the place value of decimals: tenths, hundredths and thousandths. Ask the learners to write 0.1 as a fraction ($\frac{1}{10}$). Repeat for 0.01 and 0.001. Ask them to use this information to write other decimal numbers as fractions, for example 0.4, 0.08, 0.009, 0.25, 0.125. Move on to whole numbers with decimal places, for example 2.6, 12.14, 34.275.
- Write some numbers with decimal places on the board, for example: *5.1, 5.01, 5.11, 5.001, 5.101.* Ask the learners to order these. Invite volunteers to explain how they found the order.

- Ask the learners to make up number sentences using the symbols > and <.
- Discuss how these numbers would be rounded to the nearest whole number and, where appropriate, the nearest tenth; for example 5.11 rounds to 5 because 0.11 is closer to zero than 1, or to 5.1 because 0.11 is closer to 0.1 than 0.2.
- Repeat with other numbers.
- Ask: *Where do we see decimal fractions in real life?* Aim towards money and measurements. Ask the learners to write down amounts of money, for example $12.75, and to explain why the 75 is 75 hundredths.
- Repeat with measures, for example 5.5 cm, 10.25 m, 6.175 kg, 5.1 litres.
- Ask the learners to complete photocopiable page 139.

Plenary

- Ask the learners to describe to a partner what they have been learning about during the lesson.
- Take feedback from the photocopiable activity. Invite individuals to share the numbers they made and explain how they rounded them.

Success criteria

Ask the learners:

- What is meant by a decimal number? Can you give an example of a number with three decimal places? Two? One?
- Which of these is the largest number: 2.1, 2.01, 2.001, 2.11? How do you know?
- Where do we find decimal numbers in real life? Where else?

Ideas for differentiation

Support: During the photocopiable activity, ask these learners to use three digit cards to make up numbers with one decimal place.

Extension: During the photocopiable activity, ask these learners to use five digit cards to make up numbers with three decimal places.

Name: _____

Decimals

12.55 m, 24.76 kg, 11.45 l

You will need:

A set of 0 to 9 digit cards.

What to do

- Use four digit cards to make numbers with two decimal places. Decide whether each number is an amount in metres, kilograms or litres. It would be good to make a mixture of them all!

- Put your numbers and amounts in the table below and complete the rest of the table. An example has been done for you.

Numbers picked	Amount made	Round to the nearest tenth	Round to the nearest whole number
2, 4, 9, 3	24.93 m	24.9 m	25 m

- Now write down all your decimal numbers in order from smallest to largest.

_____

_____

Percentages 1

Learning objectives

- Understand percentage as the number of parts in every 100 and find simple percentages of quantities. (5Nn19)
- Express halves, tenths and hundredths as percentages. (5Nn20)
- Solve single and multi-step word problems (all four operations); represent them, e.g. with diagrams or a number line. (5Pt2)

Resources

Pencils; paper; photocopiable page 141; set of digit cards from photocopiable page 9 for each learner; stopwatch for each pair.

Starter

- Explain that the learners will rehearse multiples and factors.
- Call out multiples of 6, 7, 8 and 9. Ask the learners to write down their factors.
- Ask the learners to write down multiples of 6. Give them a minute. Repeat for multiples of 7, 8 and 9.

Main activities

- Ask: *What can you tell me about percentages?* Establish that a percentage is a special fraction and that 1% is $\frac{1}{100}$. Ask the learners to write down different percentages and their fraction equivalences, for example 45% = $\frac{45}{100}$.
- Give the learners this statement: *Ali has $200. It is 100% of his money. Can you work out 30% of his money?* Demonstrate how to use 10% to find this: 10% is $200 ÷ 10 which is $20, so 30% is three times that, $60. Ask them to work out as many other percentages as they can by finding 10% then doubling, halving, adding and subtracting. For example 10% is $20 so 5% is $10, $2\frac{1}{2}$% is $5, 20% is $40, 15% is $30.
- Repeat for $340.

- Ask the learners to discuss and solve this problem: *There were 340 people at the fair. 40% of them were adults. The rest were children. How many were children?* Agree that finding 10% first is sensible. You can then double and double again to find 40%, which is 136. Therefore there were 136 adults. Take that from 340. So 204 must be children.
- Repeat with similar problems.
- Ask the learners to complete photocopiable page 141.

Plenary

- Take feedback from the photocopiable activity. Invite individuals to share the number they made up and the percentages they found.
- Ask the learners to tell the class one of their problems and ask them to solve it.
- Finish the lesson by writing up a three-digit number and asking the learners to tell you different percentages of that number.

Success criteria

Ask the learners:

- How would you describe a percentage to someone who doesn't know? How else?
- If 100% is $200, what else do we know? What else? How did you work that out?
- If there are 30 children in the class and 40% are boys, how many are girls? How do you know?

Ideas for differentiation

Support: These learners should focus on finding percentages which are multiples of 10%, for example 20%, 40%.

Extension: Challenge these learners to explore more complicated percentages, for example $2\frac{1}{2}$%, 24%, $37\frac{1}{2}$%, 125%.

Name: _____

Finding percentages

You will need:

A partner and a set of 0 to 9 digit cards.

What to do

- Pick three digit cards and use them to make a three-digit number.

- Write down as many percentages of that number as you can in the space below. Time yourself for two minutes.

_____

_____

_____

_____

- Take some of these percentages and make up three problems using them.

Problem 1

_____

_____

_____

_____

Problem 2

_____

_____

_____

_____

Problem 3

_____

_____

_____

_____

Percentages 2

- Know by heart pairs of one-place decimals with a total of 1, e.g. 0.8 + 0.2. (5Nc1)
- Derive quickly pairs of decimals with a total of 10, and with a total of 1. (5Nc2)
- Understand percentage as the number of parts in every 100 and find simple percentages of quantities. (5Nn19)
- Solve single and multi-step word problems (all four operations): represent them, e.g. with diagrams or a number line. (5Pt2)

Pendulum (three interlocking cubes on a piece of string or similar); pencils; paper; photocopiable page 143.

Starter

- Tell the learners they will be practising finding pairs of decimal numbers that total 1 and 10.
- Swing the pendulum. As it swings one way, call out one-place decimals, such as, 0.7, 0.4. As it swings the other, the learners should call out the number that goes with it to make 1, for example 0.3, 0.6.
- Repeat for decimals that total 10, for example 3.8 (6.2), 1.9 (8.1).

Main activities

- Recap that a percentage is a special fraction and that 1% is $\frac{1}{100}$. Ask the learners to write down different percentages and their fraction equivalences, for example 72% = $\frac{72}{100}$.
- Ask: *What is a good way to find percentages of different amounts?* Take feedback. Agree that finding 10% is helpful, they can then use doubling, halving, adding, subtracting and multiplying to find others.
- Give the learners this statement: *100% is equivalent to $180.* Ask them to work out as many other percentages as they can from this statement on paper. Take feedback on the percentages they find. They should explain their strategies. Repeat for other amounts.

- Ask the learners to discuss and solve this problem: *Jeans costing $45 were in a sale. There was a 20% discount. What is the new price of the jeans?* Agree that 10% of $45 is $4.50 ($45 ÷ 10), so 20% must be double that ($9). $9 taken from $45 is $36, the new price of the jeans.
- Repeat with similar problems.
- Ask the learners to complete photocopiable page 143.

Plenary

- Ask the learners to find different percentages of different amounts of money, for example 15% of $50. Each time expect volunteers to explain how they worked out the answer.
- Ask them to assess their understanding of what percentages are and how to find them.

Ask the learners:

- How could you work out 15% of a number? How else?
- If 100% is $150, what else do we know? What else? How did you work that out?
- There are 120 loaves of bread in a shop. 45% are past their sell by date. How many are not? How do you know?

Support: During the photocopiable activity, ask these learners to make two-digit even numbers and find percentages which are multiples of 10%, for example 20%, 40%.

Extension: During the photocopiable activity, challenge these learners to find more complicated percentages, for example $2\frac{1}{2}$%, 24%, $37\frac{1}{2}$%, 125%. You could alter the central amount in the percentage web to make it more complex.

Name: _____

Percentage web

1. Imagine 100% is $360. Write down other percentages of this amount. See how many you can find in two minutes!

   Some have been done for you.

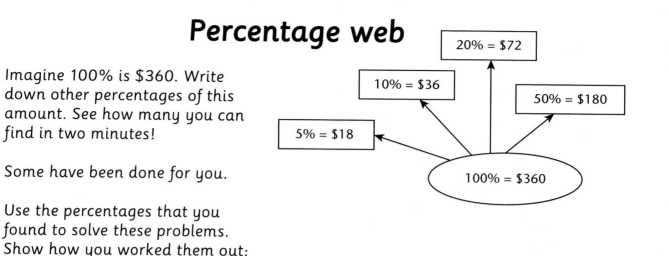

2. Use the percentages that you found to solve these problems. Show how you worked them out:

a) My friend bought a DVD player in the winter sales. Its original cost was $360. It had been reduced by 35%. How much did she pay?

_____

_____

_____

b) Another friend bought an expensive coat in the sale. It was $360. In one shop it had been reduced by 64%. In another shop it had been reduced by 72%. What is the difference in the new prices of the coat?

_____

_____

_____

c) There were 720 people at the carnival.
   Half of them were children.
   Of the children 35% were girls and the rest were boys.
   How many boys and how many girls were at the carnival?

_____

_____

_____

Equivalences 1

- Express halves, tenths and hundredths as percentages. (5Nn20)
- Solve single and multi-step word problems (all four operations): represent them, e.g. with diagrams or a number line. (5Pt2)
- Solve a larger problem by breaking it down into sub-problems or represent it using diagrams. (5Ps10)

Resources

Calculators; pencils; paper; photocopiable page 145 for each pair or group; scissors.

Starter

- Explain that the learners will rehearse place value.
- Give each learner a calculator. Ask them to key in 34. Then give instructions, such as: *Put 1 in front of the 3, change the 3 to a 5.*
- For each instruction, the learners should work out what to do to change the number, for example to put 1 in front of the 3 they add 100, to change 3 to 5 they add 20.
- Repeat several times.

Main activities

- Ask: *What is a way of expressing 1% as a fraction?* Agree $\frac{1}{100}$. Repeat for 50%. Ask: *Can you think of two ways of expressing 10% as a fraction?* Agree $\frac{10}{100}$ and $\frac{1}{10}$.
- Ask the learners to write down some percentages and then convert them to fractions. Take feedback. Make a list of some of those they say.
- Ask: *How would you express $\frac{4}{10}$ as a percentage?* Agree 40%. Repeat for other tenths. Move on to hundredths, for example $\frac{25}{100}$, $\frac{8}{100}$.
- Ask the learners to write some tenths and hundredths and then convert them to percentages. Take feedback, listing some of them.

- Set this problem: *There was a sale of boots in two shops. In one there was a reduction of $\frac{4}{10}$. In the other there was a reduction of 35%. Which was the largest reduction?* Ask the learners to discuss how they can work this out without knowing the price. Establish that they can convert the fraction to a percentage and compare. Agree that $\frac{4}{10}$ is 40%, which is a larger discount than 35%.
- Repeat with similar problems.
- Ask the learners to play the game on photocopiable page 145.

Plenary

- Take feedback from the game on the photocopiable activity. Invite the learners to share the pairs they found.
- Recap how the learners can change the fractions they have been working with to percentages.

Success criteria

Ask the learners:

- Why are $\frac{1}{2}$ and 50% equivalent?
- Can you tell me any other fractions that are the same as percentages? Are there any others?
- How are fractions and percentages the same? How are they different?
- Which is the least, $\frac{1}{10}$ or 1%? Why?

Ideas for differentiation

Support: Ask these learners to play the game with the tenths and multiples of 10% cards.

Extension: After they have played the game, challenge these learners to work out some other equivalences, for example $\frac{1}{4}$ and 25%. Ask them to make cards for these to add to the game.

Equivalent!

The aim of this game is to find fractions and decimals that match, for example $\frac{2}{10}$ and 20%. Play it with a partner or in a small group.

1. Cut out the cards below. Shuffle them and place them face down on the table.

2. Take it in turns to pick two cards. Look at them. If they match keep them. If they don't match, put them back in the same place you got them from. Watch carefully and try to remember where the cards have been put back!

3. The winner is the player who makes the most pairs.

$\frac{1}{10}$	$\frac{2}{10}$	$\frac{3}{10}$	$\frac{4}{10}$	$\frac{5}{10}$
$\frac{6}{10}$	$\frac{7}{10}$	$\frac{8}{10}$	$\frac{9}{10}$	$\frac{1}{100}$
$\frac{4}{100}$	$\frac{5}{100}$	$\frac{8}{100}$	$\frac{14}{100}$	$\frac{16}{100}$
$\frac{25}{100}$	$\frac{35}{100}$	$\frac{48}{100}$	$\frac{84}{100}$	$\frac{95}{100}$
10%	20%	30%	40%	50%
60%	70%	80%	90%	1%
4%	5%	8%	14%	16%
25%	35%	48%	84%	95%

Equivalences 2

Learning objectives

- Calculate differences between near multiples of 1000, e.g. 5026 – 4998, or near multiples of 1, e.g. 3.2 – 2.6. (5Nc11)
- Recognise equivalence between the decimal and fraction forms of halves, tenths and hundredths and use this to help order fractions, e.g. 0.6 is more than 50% and less than $\frac{7}{10}$. (5Nn16)

Resources

Pencils; paper; photocopiable page 147 for each pair or group; scissors.

Starter

- Write this calculation on the board: *2054 – 1998*. Ask the learners to find the solution. Invite individuals to share strategies. Congratulate any that count up from 1998 to 2054.
- Ask a variety of subtraction calculations that can be easily solved using this strategy, for example 1004 – 996, 3115 – 2998.
- Differentiate your questions according to the different attainment levels in your class.

Main activities

- Write $\frac{1}{2}$ on the board. Ask the learners to tell you another way to express this. Encourage equivalences such as $\frac{4}{8}$ and $\frac{5}{10}$, decimal notation and percentages.
- If they cannot tell you what $\frac{1}{2}$ is as a decimal, then link this to money (50 cents = $0.50) and measures (5 mm = 0.5 cm, 50 cm = 0.5 m).
- Repeat for tenths and hundredths, linking to money and measures as before, for example $\frac{1}{10}$ = 0.1 ($0.10 and 0.1 m), $\frac{1}{100}$ = 0.01 ($0.01 and 0.01 m).
- Ask the learners to discuss in pairs where they might see 50% in real life and what it means, for example half price in a sale. Discuss the fact that 50% can be recorded as $\frac{1}{2}$ or 0.5.

- Ask the learners to write down other equivalent fractions, decimals and percentages, for example $\frac{1}{10}$ = 10% = 0.1, $\frac{7}{10}$ = 70% = 0.7, $\frac{5}{100}$ = 5% = 0.05. Invite individuals to share some of the ones they made up.
- Ask the learners to prove that 0.6 is more than 50% and less than $\frac{7}{10}$.
- Organise the learners into pairs or small groups and ask them to play the game on photocopiable page 147.

Plenary

- Write sets of fractions, decimals and percentages on the board, for example 0.8, 60%, $\frac{7}{10}$. Ask the learners to order them from smallest to largest. Invite individuals to explain how to order, for example 60% (equivalent to $\frac{6}{10}$), $\frac{7}{10}$, 0.8 (equivalent to $\frac{8}{10}$).

Success criteria

- How else can you write $\frac{5}{10}$? Is there another way?
- How else can you write 25%? Is there another way?
- Where do we see decimals in real life? Where else? What about percentages?
- Which is the largest: $\frac{8}{10}$, 15% or 0.3? How do you know?

Ideas for differentiation

Support: Ask these learners to focus on tenths and link these to percentages and decimals, for example $\frac{3}{10}$ = 0.3 = 30%.

Extension: During the photocopiable activity, challenge these learners to explore other equivalences, for example $\frac{3}{4}$ = 0.75 = 75%, $\frac{5}{8}$ = 0.625 = $62\frac{1}{2}$%, $\frac{2}{5}$ = 0.4 = 40% and make domino cards for these.

Domino equivalences

1. Cut out the cards and shuffle them well. Deal seven cards to each player and place the rest face down in a pile on the table.

2. The first player turns over the top card of the pile. Look at your cards to see if you can match one to one end of the card placed face up. If you can, put it in position. If you can't, pick up another card.

3. The next player has their turn. Play continues until all the cards have been used or play cannot continue.

4. The player who has the fewest cards left wins.

$\frac{1}{10}$	0.3	$\frac{3}{10}$	20%	$\frac{2}{10}$	0.5
50%	0.6	$\frac{6}{10}$	80%	0.8	0.2
$\frac{5}{10}$	0.9	$\frac{9}{10}$	10%	0.1	30%
$\frac{10}{10}$	0.7	$\frac{7}{10}$	1	90%	40%
$\frac{4}{10}$	60%	$\frac{8}{10}$	0.4	70%	100%

Cambridge Primary: Ready to Go Lessons for Maths Stage 5 © Hodder & Stoughton Ltd 2012

Ratio and proportion 1

Learning objectives

- Use fractions to describe and estimate a simple proportion, e.g. $\frac{1}{5}$ of the beads are yellow. (5Nn21)
- Solve single and multi-step word problems (all four operations): represent them, e.g. with diagrams or a number line. (5Pt2)
- Deduce new information from existing information to solve problems. (5Ps4)

Resources

Pencils; paper; 10- and 25-cent coins; photocopiable page 149.

Starter

- Explain that the learners will be rehearsing finding fractions of numbers to help in the lesson.
- Write 24 on the board. Call out fractions of this number, for example $\frac{1}{2}$, $\frac{1}{4}$, $\frac{3}{4}$, $\frac{2}{3}$, $\frac{5}{6}$, $\frac{3}{8}$. Ask the learners to write down their answers and show you.
- Invite individuals to share their strategies for finding these.

Main activities

- Showing appropriate coins, ask: *In my wallet I have four 10-cent coins and eight 25-cent coins. What proportion of the coins are 10-cent coins? 25-cent coins?* Discuss what is meant by 'proportion'. Take feedback. Establish that it is a part of a whole amount. The learners should then apply this to the question. Agree that the whole, in this case, is the total number of coins (12). There are four 10-cent coins, so the proportion of 10-cent coins is $\frac{4}{12}$. The proportion of 25-cent coins is $\frac{8}{12}$.
- Ask: *What mathematics is involved in proportion?* Agree on fractions.
- Repeat with similar problems.
- Ask: $\frac{3}{5}$ *of a class have packed lunches, the rest eat lunch at school. How many learners have lunch at school?* Discuss this with a partner.

- Establish that there is not enough information to answer the problem. The information shows that $\frac{2}{5}$ have lunch at school. This could be any multiple of 5. Ask the learners what additional information is needed (the number of learners in the class). Say: *There are 30 learners in the class. How many have lunch at school?* Agree that 12 have lunch at school and 18 have packed lunch.
- Ask the learners to work through photocopiable page 149.

Plenary

- Ask the learners to share their answers to some of the questions on the photocopiable activity and explain how they achieved their solutions.
- Ask them to work with a partner to make up a problem involving proportion for the class to answer.

Success criteria

Ask the learners:

- How can you explain the meaning of proportion to someone who doesn't know? Is there another way?
- A bowl contains 12 fruits. 8 are papaya. The rest are mangos. What proportion of the fruits are mangos? How do you know?
- Can you make up a problem involving proportion?

Ideas for differentiation

Support: During the photocopiable activity, ask these learners to focus on the basic proportions asked for and not the additional fraction instructions.

Extension: After they have completed the photocopiable activity, challenge these learners to make up some proportion problems of their own.

Name: _____

Proportion practice

Answer the questions below.

1. There are 10 counters. 6 are red and 4 are blue.

   What proportion are red? _____

   Write this using another fraction: _____

   What proportion are blue? _____

   Write this using another fraction: _____

2. A pizza has been divided into 12 pieces. Sami eats three pieces.

   What proportion of the pizza is left? _____

   Draw a picture to show this:

3. A bowl contains 12 fruits. There are 3 bananas, 4 apples and 5 peaches.

   What proportions are:

   a) bananas _____   Write this using another fraction: _____

   b) apples _____   Write this using another fraction: _____

   c) peaches? _____   Write this using another fraction: _____

4. There are 28 learners in a class. 16 of them are 9-year-olds. The rest are 10-year-olds.

   What proportion are 10-year-olds? Write this in two ways.

   _____

5. A pie is made with apples and cherries. There is a total of 6 kg of fruit in the pie.

   The proportion of apples is $\frac{2}{3}$. How many kilograms of cherries are there?

   _____

Ratio and proportion 2

Learning objectives

- Multiply multiples of 10 to 90, and multiples of 100 to 900, by a single-digit number. (5Nc12)
- Multiply by 19 or 21 by multiplying by 20 and adjusting. (5Nc13)
- Use ratio to solve problems, e.g. to adapt a recipe for 6 people to one for 3 or 12 people. (5Nn22)
- Deduce new information from existing information to solve problems. (5Ps4)

Resources

Pencils; paper; coloured counters; photocopiable page 151; yellow and red paints; three paintbrushes for each learner; water; plain paper; yellow and blue paints.

Starter

- Explain that the learners will be rehearsing mental calculation strategies for multiplication.
- Call out different multiples of 10 and 100 and ask the learners to multiply them by single digits, for example: *90 × 8 (9 × 8 = 72, 90 × 8 = 720), 300 × 6 (3 × 6 = 18, 300 × 6 = 1800).*
- Write this calculation on the board: *23 × 19.* The learners should work this out by multiplying 23 by 20 and taking one lot of 23 away.
- Repeat for other calculations.
- Repeat with other multiples and fractions.

Main activities

- Ask: *What is meant by the word 'ratio'?* Take feedback. Establish that a ratio shows the relative sizes of two or more values. Demonstrate using one blue counter and three red. Explain that there is one blue counter for every three red so the ratio is 1:3. Repeat the demonstration for two blue counters and five red.
- Show a different number of each colour counter and ask the learners to write down the ratio. Invite a volunteer to explain how they got their answer. Repeat a few times.

- Discuss how the quantity of counters could be changed to give the same ratio. Agree that they could double each quantity; for example, in the case of one blue for every three red, it would be two blue for every six red, or 2:6, which can be reduced to 1:3. Establish that if both quantities are multiplied by the same number there will be an equivalent ratio. Demonstrate examples using the counters.
- Ask: *When might we use ratio in real life? Discuss this with a partner.* Take feedback. Agree that ratio is used in such things as recipes, mixing paints.
- Ask the learners to work through photocopiable page 151.

Plenary

- Invite the learners to share with the class their shades of orange and the ratios they made.
- Ask several learners to order their favourite shades from lightest to darkest. The class predict the ratios used.

Success criteria

Ask the learners:

- How can you explain ratio to someone who doesn't know? Is there another way?
- What does a ratio of 1:2 mean? Can you think of another explanation?
- If a recipe for 8 pancakes required 100 g of flour, 1 egg and 250 ml of milk, how much would be needed for 24 pancakes? How did you work that out?

Ideas for differentiation

Support: During the photocopiable activity, give these learners specific ratios to make, for example 1:2, 1:3 and 1:4. They should label each blob with the ratio used.

Extension: During the photocopiable activity, once these learners have made shades of orange, challenge them to explore a variety of ratios of yellow and blue paint to make green.

Name: _____

Paint blobs

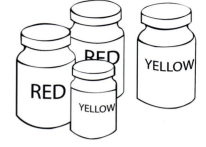

When you mix red and yellow paint you get orange.
The shade of orange depends on the ratio of the amounts
of red and yellow paint used.

You will need:

Yellow and red paint, three paintbrushes and a small
container of water.

What to do

- Experiment with the shades of orange that you can make.

- Use one of your paintbrushes to put a blob of yellow paint in the space below.
  Mix it with a blob of red the same size.

You have made orange with a ratio of 1:1 yellow to red.

- Now try with different ratios on another piece of paper.
  Start with these: 1:2, 2:3, 5:2. Then make up some of your own.

- Make a list here of the ratios that you made:

_____

_____

_____

- Look at your results. Which is your favourite shade of orange?

  Paint it here:

- What is its ratio of yellow to red?

_____

Addition and subtraction

Learning objectives

- Derive quickly pairs of decimals with a total of 10, and with a total of 1. (5Nc2)
- Find the total of more than three two- or three-digit numbers using a written method. (5Nc18)
- Add or subtract any pair of three- and / or four-digit numbers, with the same number of decimal places, including amounts of money. (5Nc19)
- Choose an appropriate strategy for a calculation and explain how they worked out the answer. (5Ps2)
- Solve a larger problem by breaking it down into sub-problems or represent it using diagrams. (5Ps10)

Resources

Pencils; paper; photocopiable page 153.

Starter

- Tell the learners they will be practising finding pairs of decimals that total 1 and 10.
- Call out a variety of decimal numbers, for example: *0.3, 0.8*. The learners should write down the numbers that go with them to make 1.
- Repeat for numbers such as *2.3, 6.2* to make 10.

Main activities

- Set this problem: *In the market Elisha spent $10.75 on meat, $9.45 on groceries and $12.05 on vegetables. How much did she spend altogether?* Ask the learners to discuss in pairs how they could find out.
- Take feedback. Agree that the amounts need adding together. Discuss possible strategies. Invite the learners to share their ideas, for example partitioning, adding the dollars first and then the cents, or sequencing, keeping one amount whole and adding the others by partitioning. Focus on the extended and, if appropriate, compact methods of addition (see the grids in the next column). Stress the importance of ensuring the decimal points line up underneath one another.

		9	.	4	5
	1	0	.	7	5
+	1	2	.	0	5
	2	0	.	0	0
	1	1	.	0	0
		1	.	1	0
			.	1	5
	3	2	.	2	5

		9	.	4	5
	1	0	.	7	5
+	1	2	.	0	5
	3	2	.	2	5
	1	1		1	

- Tell the learners that Elisha had $50. Ask them to discuss in pairs how to find how much she would have left after her visit to the market. Take feedback of their ideas. Encourage mental calculation strategies such as complementary addition and sequencing. Invite volunteers to share these on the board. If appropriate, demonstrate the compact method.
- Repeat for other scenarios involving money.
- Ask the learners to work through photocopiable page 153.

Plenary

- Invite volunteers to share the problems they made up in the photocopiable activity. Find the solutions as a class. Each time, discuss the best method to use.
- Ask the learners to assess their confidence in adding and subtracting decimal numbers.

Success criteria

Ask the learners:

- What is the total of $24.75, $13.25 and $10.99? How did you work that out? Is there another way?
- If someone spent $55.50, how could we find their change from $75? Is there another way?
- What strategies for addition and subtraction have we thought about today?

Ideas for differentiation

Support: Adapt the photocopiable activity so these learners focus on adding and subtracting two-digit numbers with one decimal place.

Extension: Adapt the photocopiable activity so these learners add four four-digit numbers with two decimal places; for example, add a fourth side to Ibrahim's field.

Problems

Ibrahim bought 75 m of fencing. He wanted to build a fence around a plot of land for his goats. He measured each side of the plot of land. One side was 15.75 m long, the second was 24.6 m long, and the third was 18.26 m. He will keep 2 m unfenced so that he can put a gate in.

How much fencing does he need?

How much fencing will he have left over?

What to do

- Decide what to do to answer to this problem.

- Decide which strategies to use.

- Think about what you need to do to the 24.6 m to make it comparable to the other measurements.

- Show your working here:

Ibrahim needs _____ of fencing.

_____ of fencing will be left.

- Make up a similar problem. Write it below and then solve it.

_____

_____

_____

_____

_____

_____

Mental calculation strategies

Learning objectives

● Describe and continue number sequences, e.g. −30, −27, *, *, −18...; identify the relationships between numbers. (5Ps6)

● Multiply by 25 by multiplying by 100 and dividing by 4. (5Nc14)

● Use factors to multiply, e.g. multiply by 3, then double to multiply by 6. (5Nc15)

● Estimate and approximate when calculating, e.g. using rounding, and check working. (5Pt6)

Resources

Pencils; paper; photocopiable page 155; set of 0 to 9 digit cards from photocopiable page 9 for each learner.

Starter

• Explain to the learners that they will be rehearsing number sequences.

• Write this on the board: *−15, −8,* ☐ *, 6,* ☐ *,* ☐ *, 27.* Ask the learners to work out the missing numbers and to extend the sequence in both directions.

• Repeat with other sequences that involve both positive and negative numbers.

Main activities

• Set this problem: *Samir, the manager of a sports equipment store, was sent 25 boxes of tennis balls. In each box there were 56 tennis balls. How many tennis balls did Samir receive altogether?* Establish that to find the answer 56 is multiplied by 25. Take feedback, inviting the learners to share all the strategies that they can think of to do this.

• Focus on the strategy of multiplying by 100 and dividing by 4 (by halving and halving again): 5600, 2800, 1400. Write similar calculations on the board for the learners to practise.

• Ask: *Jodie and her 5 friends each had 26 toffees. How many did they have altogether?* Ask the learners to think of the ways that they could multiply 26 by 6.

• Discuss using factors to multiply. Ask them to multiply the 26 by 3 and then double. Explain that this works because 3 × 2 = 6.

• Make up similar problems for the learners to answer. These should include using other factors, such as multiplying by 8 by multiplying by 4 and doubling, and by 9 by multiplying by 3 twice.

• Ask the learners to work through photocopiable page 155.

Plenary

• Invite the learners to share with the class some of the numbers they made and the answers when multiplied by 25 and by 6.

• Invite volunteers to demonstrate how to find the answer on the board.

• Make up problems similar to those in the main part of the lesson for the learners to solve.

Success criteria

Ask the learners:

● Can you explain how to multiply by 25 by multiplying by 100 and dividing by 4 to someone who doesn't know? Why does this strategy work?

● Can you give me an example of how we can use factors to help us multiply? How does this strategy work? Are there any others?

● What is 78 × 25? How did you work that out?

Ideas for differentiation

Support: During the photocopiable activity, these learners should use the digit cards 1, 2, 3, 4 and 5 to make their numbers.

Extension: During the photocopiable activity, challenge these learners to make three-digit numbers to multiply.

Name: _____

Which strategy?

You will need:

Digit cards 3, 4, 5, 6, 7, 8 and 9.

What to do

- Shuffle the digit cards, pick two and use them to make a two-digit number.

- Multiply your number by 25 using the strategy multiply by 100 and divide by 4.

- Show your working and answer in the table below.

- Now multiply your number by 6. Use the strategy of multiplying by 3 and doubling.

- Show your working and answer in the table below.

    An example has been done for you.

Number	×25	Answer	×6	Answer
93	93 × 100 = 9300 $\frac{1}{2}$ of 9300 = 4650 $\frac{1}{2}$ of 4650 = 2325	2325	93 × 3 = 279 279 × 2 = 558	558

Multiplication and division

Learning objectives

- Multiply two-digit numbers with one decimal place by single-digit numbers, e.g. 3.6 × 7. (5Nc22)
- Start expressing remainders as a fraction of the divisor when dividing two-digit numbers by single-digit numbers. (5Nc24)

Resources

Pencils; paper; photocopiable page 157; set of digit cards from photocopiable page 9 for each learner.

Starter

- Tell the learners that they will practise multiplying two-digit decimal numbers by single-digit numbers.
- Write some three-digit decimal numbers on the board, for example 13.5, 24.7, 19.3. Call out a single-digit number, for example 4. The learners should multiply the numbers on the board by this.
- Repeat using the same three-digit numbers but a different single-digit number.

Main activities

- Set this problem: *Maggie received a supply of apples to sell in her shop. 3 crates were delivered. In each crate were 125 apples. How many apples were there altogether?* Ask the learners to discuss this in pairs and then work out the answer. Take feedback on the strategies used, for example partitioning, grid method, compact method. See appropriate lessons in Units 1 and 2 for details. Agree that Maggie had 375 apples.
- Say: *Maggie decided to put them all in bags of 8. How many bags did she make up?* Again ask the learners to work in pairs to find out. Agree that she would have 46 bags with 7 apples left over. Ask: *How could we express this remainder?* Take feedback. If not suggested, lead them to think in terms of fractions: 7 out of the 8 left, so the remainder could be written as $\frac{7}{8}$.

- Repeat for similar scenarios.
- Ask the learners to work through photocopiable page 157.

Plenary

- Ask: *What is the word for the answer to a division?* Establish that it is quotient. Invite the learners to share the quotients they made during the photocopiable activity and the fractions they used to express the remainder.
- Ask: *How confident are you at expressing a remainder as a fraction?*

Success criteria

Ask the learners:

- How would you describe how to express a remainder as a fraction to someone who doesn't know? Is there another way?
- Maggie decided to put the 375 apples into bags of 10. How many bags could she fill? What would the remainder be, expressed as a fraction?
- Can you make up a problem that involves multiplying and dividing?

Ideas for differentiation

Support: During the photocopiable activity, these learners should make two-digit numbers to divide by numbers up to and including 5.

Extension: During the photocopiable activity, these learners should divide by a two-digit number between 11 and 30.

Name: _____

Remainders

You will need:

A partner, two sets of 0 to 9 digit cards, paper and a pencil.

What to do

- Shuffle both sets of digit cards and place them in two piles on the table.

- Take it in turns to:

    a) Pick three digit cards from one pile and make up a three-digit number.

    b) Take a card from the second pile.

    c) Divide the three-digit number by the number on the other card.

- Use a separate piece of paper to work out your answers.

    You score:

    - 0 points if there isn't a remainder

    - 1 point if the remainder is less than half

    - 2 points if the remainder is half or more.

- The winner is the player with the most points.

- Complete the table below. An example has been done for you:

Name:				Name:			
Calculation made	Answer	Fraction	Points	Calculation made	Answer	Fraction	Points
358 ÷ 4	89 r2	$\frac{2}{4}$ or $\frac{1}{2}$	2	146 ÷ 6	24 r2	$\frac{2}{6}$ or $\frac{1}{3}$	1

Brackets

- Double any number up to 100 and halve numbers to 200 and use this to double and halve numbers with one or two decimal places, e.g. double 3.4 and half of 8.6. (5Nc16)
- Double multiples of 10 to 1000 and multiples of 100 to 10 000, e.g. double 360 or double 3600, and derive the corresponding halves. (5Nc17)
- Begin to use brackets to order operations and understand the relationship between the four operations and how the laws of arithmetic apply to multiplication. (5Nc27)
- Deduce new information from existing information to solve problems. (5Ps4)
- Explain methods and justify reasoning orally and in writing; make hypotheses and test them out. (5Ps9)

Resources

Pendulum (three interlocking cubes on a piece of string or similar); pencils; paper; photocopiable page 159.

Starter

- Explain that the learners will rehearse doubling and halving two-digit numbers.
- Swing the pendulum from side to side. As it swings one way you call out a number up to 100. As it swings the other way the learners should call out the double of that number.
- Stop occasionally to ask the learners what strategies they used to find the double.
- Repeat for halves of even numbers to 200 and doubles and halves of multiples of 10 to 1000.

Main activities

- Write this calculation on the board: $14 + 6 \times 10$. Say: *I think the answer to this is 74. What do you think?* Establish that you multiplied 6 by 10 and added 14. Ask the learners to think of the other solution. Agree that they could add 14 and 6 then multiply by 10 (200).

- Ask the learners what they think might help them to know which order to work through this to find the solution. If they do not know, establish that brackets are needed. The contents of brackets are always worked out first. So the calculation should have been written $14 + (6 \times 10)$.
- Write this on the board: $12 \times 3 + 20 \div 4$. Agree that the answers could be, for example, 14 and 41. Put in brackets: $(12 \times 3) + (20 \div 4)$. Agree that the answer is 41.
- Repeat for other calculations.
- Ask the learners to work through photocopiable page 159.

Plenary

- Invite the learners to share some of the calculations they made in the photocopiable activity. The rest of the class should find two possible solutions.
- Write this on the board: $12 \times 5 - 15 + 7 \times 3$. Invite the learners to suggest possible positions for the brackets.

Success criteria

Ask the learners:

- Why is it sometimes necessary to use brackets in a calculation?
- How would you explain how to use brackets to someone who doesn't know? How else?
- Can you make up an example of a calculation that can have two different answers? Can you explain where the brackets would need to go to give one of those answers?

Ideas for differentiation

Support: During the photocopiable activity, ask these learners to focus on calculations with one set of brackets, for example $10 \times 5 - 3$, $10 \times (5 - 3)$.

Extension: During the photocopiable activity, challenge these learners to make up calculations involving all four operations and three sets of brackets.

Name: _____

Brackets

Answer the following questions and record your work in the table below. When you make your calculations, try to include a mixture of operations.

1. Make up a calculation that involves two operations, for example 12 × 3 + 6.

   Work out the answer, working from left to right, for example 12 × 3 = 36, 36 + 6 = 42.

2. Next add brackets so that you get a different answer, for example 12 × (3 + 6) = 12 × 9 = 108.

3. Repeat steps 1 and 2 two more times.

4. Now make up three calculations that involve three operations, for example 36 ÷ 9 + 5 × 2.

   Work out the answer, working from left to right, for example 36 ÷ 9 = 4, 4 + 5 = 9, 9 × 2 = 18.

5. Next add brackets, for example (36 ÷ 9) + (5 × 2).

   Work out the new answer, for example (36 ÷ 9) = 4, (5 × 2) = 10, 4 + 10 = 14.

6. Repeat steps 4 and 5 two more times.

Calculation	1st solution	With brackets	2nd solution
12 × 3 + 6	42	12 × (3 + 6)	108

Unit assessment

- What is meant by a fraction?
- How would you work out $\frac{5}{8}$ of 48?
- What is a percentage?
- What can you tell me about ratio?

- There are 30 learners in a class. 18 of them are aged ten. Explain how you can work out what proportion are 10-year-olds.

Summative assessment activities

Observe the learners while they take part in these activities. You will quickly be able to identify those who appear to be confident and those who may need additional support.

Fractions

This activity assesses the learners' knowledge of fractions.

You will need:

Strips of paper the same length; paper; pencils.

What to do

- Organise the learners into groups of four.
- Give them some strips of paper.
- Ask them to fold the strips to make halves, quarters, eighths, thirds and sixths.
- Next ask them to use these strips and write down some equivalent fractions.
- Ask them to order the fractions from smallest to largest.

Fractions and percentages

This activity assesses the learners' knowledge of equivalences between fractions and percentages.

You will need:

Paper; pencils; cards from photocopiable page 145.

What to do

- Organise the learners into groups of two or three.
- Lay the fraction and percentage cards from page 145 face up on the table.
- Ask the learners to take it in turns to take two cards that show the same value.

Distribute copies of photocopiable page 161. Ask the learners to read the questions and write the answers. They should work independently.

Working with numbers 3

50¢ 10¢ 10¢ 5¢ 5¢
1¢ 1¢ 1¢ 1¢ 1¢

1. Write down four fractions that are equivalent to $\frac{3}{6}$.

_____

2. Work out the following:

a) $\frac{3}{4}$ of 24 _____

b) $\frac{3}{5}$ of 50 _____

c) $\frac{5}{8}$ of 32 _____

3. Order these fractions from smallest to greatest:

$\frac{1}{5}$ $\frac{1}{2}$ $\frac{1}{8}$ $\frac{1}{3}$ $\frac{1}{6}$ _____

4. Write these as decimals:

a) $4\frac{1}{10}$ _____   b) $5\frac{7}{10}$ _____   c) $3\frac{4}{100}$ _____

d) $2\frac{1}{5}$ _____   e) $3\frac{1}{2}$ _____

5. If 100% is equivalent to $120, write down five other percentages of this amount.

_____

6. The ratio of red counters to blue counters is 3 : 4. There are 9 red counters. How many blue counters are there?

_____

7. Pru had collected 76 bags of coins. There were 24 coins in each bag. How many coins had she collected altogether?

8. There are 136 books that need to be put on some shelves. 9 books can go on each shelf. How many shelves will be filled?

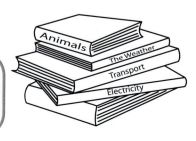

Triangles

Learning objectives

- Identify and describe properties of triangles and classify as isosceles, equilateral or scalene. (5Gs1)
- Identify simple relationships between shapes, e.g. these triangles are all isosceles because ... . (5Ps7)

Resources

Pendulum (three interlocking cubes on a piece of string or similar); pencils; paper; photocopiable page 163; rulers; set squares.

Starter

- Explain that the learners will rehearse doubling and halving.
- Swing the pendulum. As it swings one way, you call out a number to 100. As it swings the other way, the learners should call out its double. Repeat for decimal numbers.
- Repeat for halves of even numbers to 200 and 'even' decimal numbers.

Main activities

- Ask the learners to sketch six different triangles. Ask them to show any equilateral triangles and describe their properties: three equal sides, three equal angles, three lines of symmetry.
- Repeat for isosceles triangles and scalene triangles.
- Ask the learners if they have drawn any triangles with a right angle. If so, ask them to name the triangle. Establish that isosceles and scalene triangles can have a right angle.
- Say: *Sandy drew a triangle with two right angles. Is this possible?* Agree that it is not because two right angles total 180°, the same as the total of the angles inside a triangle.
- Say: *Tim drew an isosceles triangle with an obtuse angle. Is this possible?* Invite a volunteer to show that it is.
- Repeat with similar statements.

- Draw a Carroll diagram on the board similar to this:

	Right angles	No right angles
Obtuse angles		
No obtuse angles		

Invite individuals to sketch triangles in the appropriate places. Ask: *Which is the only place an equilateral triangle can be placed?* Agree on the bottom right-hand corner.
- Repeat with a different Carroll diagram, asking the learners to suggest the headings.
- Ask the learners to work through photocopiable page 163.

Plenary

- Invite volunteers to sketch one of the triangles they drew during the photocopiable activity.
- The class identifies which type each one is, naming the angles and, if appropriate, where the lines of symmetry go.

Success criteria

Ask the learners:

- How would you describe an equilateral triangle? Is there another way?
- Why do all isosceles triangles have one line of symmetry?
- Can a triangle have a right angle and an obtuse angle? Why not?
- Can a triangle have three acute angles? Why?

Ideas for differentiation

Support: During the photocopiable activity, ask these learners to focus on one of each type of triangle.

Extension: During the photocopiable activity, challenge these learners to estimate the size of each angle and add this to their descriptions.

Triangles

You will need:

A ruler, a set square (for making right angles) and a pencil.

What to do

- Draw the following triangles, then describe them, including their types of angle (acute, obtuse, right) and number of lines of symmetry.

a) An isosceles triangle with 3 acute angles

Describe your triangle:

_____
_____
_____
_____

b) An isosceles triangle with a right angle

Describe your triangle:

_____
_____
_____

c) An equilateral triangle

Describe your triangle:

_____
_____
_____
_____

d) A scalene triangle with a right angle

Describe your triangle:

_____
_____
_____
_____

e) A scalene triangle with an obtuse angle

Describe your triangle:

_____
_____
_____
_____

- Now mark the lines of symmetry on the triangles that have them.

Polygons

Learning objectives

- Create patterns with two lines of symmetry, e.g. on a pegboard or squared paper. (5Gs3)
- Recognise reflective and rotational symmetry in regular polygons. (5Gs2)
- Predict where a polygon will be after reflection where the mirror line is parallel to one of the sides, including where the line is oblique. (5Gp2)
- Explain methods and justify reasoning orally and in writing; make hypotheses and test them out. (5Ps9)

Resources

Squared paper; coloured pencils; pencils; paper for recording; mirrors; photocopiable page 165; plain paper; rulers; scissors.

Starter

- Explain that the learners will make symmetrical patterns to help them during the lesson.
- Give each learner a piece of squared paper. Ask them to fold it in half twice to make four quadrants.
- Ask the learners to colour in 15 squares in the top-left quadrant and pass their pattern to a partner. Their partner then colours the pattern in the top-right quadrant so that it is symmetrical.
- They do this twice more so that the paper has a symmetrical pattern in all four quadrants.

Main activities

- Ask the learners to sketch a variety of regular and irregular polygons (shapes with three or more straight sides). Invite individuals to draw those they sketched on the board and to name them.
- Discuss the properties of each of the shapes. Include: number of sides and corners (whether they are equal or not), types of angle (acute, obtuse, right) and whether there are any lines of symmetry. Invite volunteers to draw on the lines of symmetry where appropriate.
- Ask the learners to draw a square. Tell them that there is a mirror line against the right side. Ask them to draw the reflection. Demonstrate on the board.

- Next ask them to visualise where the square would be after a clockwise rotation of 90° around one corner and draw it. If necessary remind them what a rotation is and ask them to stand and make rotations of different sizes.
- Repeat this for a variety of simple polygons, giving the position of the mirror line each time. Invite individuals to share their results.
- Sketch this irregular pentagon on the board and a diagonal mirror line:

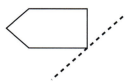

Ask: *Can you predict where the reflected shape will be? Draw what you think.* Demonstrate how to use a mirror to check and then draw the reflection. Did they predict correctly?
- Ask the learners to work through photocopiable page 165.

Plenary

- Invite the learners to share their investigations from the photocopiable activity. For each investigation ask one pair to describe what they did and the results they achieved.
- The rest of the class should compare what they hear with their own results.

Success criteria

Ask the learners:

- Which polygons can you name? Are there any others? Can you describe their properties?
- How would you describe whether a shape is symmetrical or not? Is there another way?
- What can you tell me about the number of lines of symmetry in regular shapes?

Ideas for differentiation

Support: During the photocopiable activity, ask these learners to choose one investigation to explore.

Extension: When these learners have completed the photocopiable activity, challenge them to devise an investigation of their own which involves rotation, working in pairs.

Name: _____

Playing around with polygons

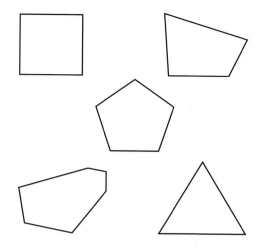

You will need:

A partner, scissors, squared paper, plain paper, a pencil, a ruler and a mirror.

What to do

● Decide which investigation you will do first. Together, do one at a time. Compare your results.

Investigation 1

1. Draw a rectangle on a piece of squared paper.

2. Draw diagonal mirror line. It should look like this:

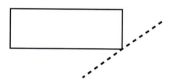

3. Now draw where you predict the reflection will be.

4. Check your prediction with a mirror.

5. Experiment with other regular and irregular polygons, for example triangle, square and hexagon.

Investigation 2

1. Draw a square on a piece of squared paper and cut it out.

2. Draw round it on a piece of plain paper.

3. Rotate the square and draw round it again. Do this several times.

4. What pattern did you make?

5. Try again with other polygons.

Perpendicular and parallel lines

Learning objectives

- Recognise perpendicular and parallel lines in 2D shapes, drawings and the environment. (5Gs5)
- Deduce new information from existing information to solve problems. (5Ps4)

Resources

Set of follow-me cards from photocopiable page 11; pencils; paper; photocopiable page 167; rulers; coloured pencils.

Starter

- Explain that the learners will practise their multiplication and division facts.
- Share out the follow-me cards between small groups. Keep one card. Read out the multiplication, for example: *4 × 6*. The group with the answer calls it out and reads the multiplication on their card. Continue. You will have the last answer.
- Start with the answer on your card. The group that has the multiplication question that goes with it reads it out, then reads out the next answer and so on.

Main activities

- Ask: *What is meant by the term 'parallel lines'?* Agree that these are two lines that never meet. They are always the same distance apart. Ask the learners to show what they look like using their arms.
- Ask them to look around the classroom and identify different pairs of parallel lines, for example edges of book covers, window frames. Give them two minutes to make a list, then ask them to swap lists with a partner. The partner tries to identify those they have listed. They feed these back to the class.
- Ask: *What is meant by the term 'perpendicular'?* Agree that these are pairs of lines that meet at right angles. Repeat the activities above but for perpendicular sides.

- Ask the learners to sketch a square, a rectangle, and some regular and irregular pentagons and hexagons. Ask them to identify those with parallel sides and those with perpendicular sides. Take feedback, inviting individuals to draw some of these shapes on the board and point out the two types of lines.
- Ask the learners to work through the activity on photocopiable page 167.

Plenary

- Invite the learners to share the patterns that they made up in the photocopiable activity. The class should identify the shapes that can be seen in the patterns.
- Recap what is meant by parallel and perpendicular lines and where they can be seen.

Success criteria

Ask the learners:

- Can you explain what is meant by parallel lines to someone who doesn't know? Can you explain in a different way?
- Can you explain what is meant by perpendicular lines? Is there another way?
- Where can we see parallel lines in real life? What about perpendicular lines?

Ideas for differentiation

Support: During the photocopiable activity, ask these learners to draw two pairs of parallel and perpendicular lines.

Extension: During the photocopiable activity, ask these learners to work out ways to create quadrilaterals in their pattern and to name them using the correct mathematical vocabulary.

Name: _____

Parallel and perpendicular

You will need:

A ruler, a pencil and some coloured pencils.

What to do

● Make up an abstract pattern using four pairs of parallel lines and four sets of perpendicular lines.

Here is the start of one:

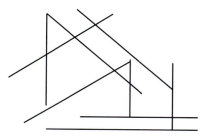

● Draw your pattern in the space below:

● What shapes can you see in your pattern? List them here:

_____

_____

● Colour your shapes in different colours. Choose one colour for each shape.

3D shapes

- Visualise 3D shapes from 2D drawings and nets, e.g. different nets of an open or closed cube. (5Gs4)
- Deduce new information from existing information to solve problems. (5Ps4)
- Recognise the relationships between different 2D and 3D shapes, e.g. a face of a cube is a square. (5Pt5)

Interlocking cubes: paper; sticky tape; scissors; photocopiable page 169.

Starter

- Explain that the learners will practise doubling and halving.
- Call out multiples of 10 to 1000, for example: *360*. Once you say the number clap. After the clap the learners should call out the double of that number.
- Repeat for multiples of 100 to 10 000, for example 3600.

Main activities

- Ask: *What 3D shapes do you know?* Take feedback, listing those they say on the board. Ensure these are included: tetrahedron, square-based pyramid, cube, cuboid, cylinder, sphere, cone, triangular prism, pentagonal prism, hexagonal prism.
- Recap the vocabulary of 3D shapes including 'faces', 'edges', 'vertices', 'prisms'.
- Invite individuals to pick a shape from the list and give clues to help the class identify it. After each clue, the class has to guess what the shape is.
- Give each learner a cube, a piece of paper, scissors and sticky tape. Ask them to look carefully at the cube and imagine what it would look like opened up, then sketch what they visualised, cut it out, and fold and stick it to make a cube.

- Tell the learners that they created a net of a cube. Ask them to imagine a triangular prism. Draw one on the board to help them:

Again, ask the learners to sketch what they think the net would look like, cut it out, and fold and stick to make the shape. Compare the nets they made by asking volunteers to draw theirs on the board.
- Ask the learners to work though photocopiable page 169.

Plenary

- Invite pairs to share what they did in the photocopiable activity. Ask them to show the shape they originally made; the class then estimates how many cubes are needed to make the cube / cuboid.
- Discuss where cubes and cuboids are found in real life.

Ask the learners:
- Can you describe how to make a net of a cube? Is there another way?
- How would you make a triangular prism? Could you do it in a different way?
- What is a prism? Can you give an example? Any others?
- Where do we see cubes and cuboids in real life? Where else?

Support: If these learners have difficulty visualising the net of a cube, allow them to draw around the faces, cut them out and stick them together to make the cube.

Extension: Challenge these learners to work out two ways to make a net for a cube.

Name: _____

Making 3D shapes

You will need:

A partner and some interlocking cubes.

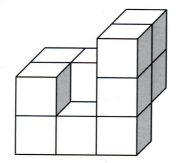

What to do

- Take it in turns to build a shape with 10 cubes. Here is an example:

- Once you have made a shape, give it to your partner. Your partner estimates the smallest number of cubes that are needed to turn your shape into a cube.

- Try this out. How close was your partner's estimate to the actual number of cubes needed?

- Record your results in this table.

Name:		
Estimated number of cubes needed	**Actual number of cubes needed**	**Difference**
Name:		
Estimated number of cubes needed	**Actual number of cubes needed**	**Difference**

- Do this again, but this time make a cuboid. Make your own chart on the back of this paper to record your work.

Transformations

Learning objectives

- Read and plot co-ordinates in the first quadrant. (5Gp1)
- Understand translation as movement along a straight line, identify where polygons will be after a translation and give instructions for translating shapes. (5Gp3)
- Deduce new information from existing information to solve problems. (5Ps4)

Resources

Squared paper; pencils; rulers; counters; photocopiable page 171, card, scissors, colouring pencils.

Starter

- Explain that the learners will rehearse reading and plotting co-ordinates.
- Give each learner a piece of squared paper. Ask them to draw a vertical axis on the left and a horizontal axis perpendicular to the vertical one. They label these from 1 to 10 with 0 where the lines meet.
- Call out co-ordinates for them to plot, for example: *(4, 6)*. Do this so that they make various polygons.

Main activities

- Ask: *What is a translation? Discuss this with a partner.* Take feedback. Agree that a translation is a movement from one place to another. Ask the learners to give examples of translations that they may see in real life, for example people walking from one place to another, patterns on a mosaic.
- Give each learner a counter and piece of squared paper. Ask them to place the counter in the top-left square. Call out instructions, for example: *Move the counter 4 squares to the right, 3 squares down, 2 left, 1 up.* When you have given your instructions the learners should compare the positions of their counters. Are they all in the same position? Repeat this a few times.

- Ask the learners to do the above activity with a partner. One gives the instructions, keeping a note of them by marking the movement on their paper. The other moves their counter according to the instructions. When finished, they compare finishing points to ensure the 'mover' followed the instructions correctly. They then swap roles.
- Invite pairs to demonstrate what they did by giving their instructions for the class to follow.
- Ask the learners to work through photocopiable page 171.

Plenary

- Ask the learners to demonstrate the patterns they made in the photocopiable activity by drawing their pattern on the board and describing the movements made.
- Ask the learners to assess their confidence in making translations.

Success criteria

Ask the learners:

- How could you describe translation to someone who doesn't know? How else?
- Where do translations occur in real life (for example wallpaper, keys on a piano)? Can you think of anywhere else?
- How could I move the square on the board to this position? Is there another way?

Ideas for differentiation

Support: During the photocopiable activity, ask these learners to create their pattern on squared paper.

Extension: During the photocopiable activity, ask these learners to explore making a pattern with translations and also some rotations. These should be rotations of 90°.

Name: _____

Translating patterns

You will need:

Card, a pencil, a ruler, scissors and colouring pencils.

What to do

- Choose a polygon to make a translating pattern. Draw it on a piece of card and cut it out.

- Draw round your polygon in the space below, then translate it.
  Use a ruler to measure the length of your translation. Draw around
  your polygon again. Keep doing this until you have a pattern.

- As you work, make a note of the translations you make.

- Now colour your pattern.

Here is an example of the beginning of a pattern:

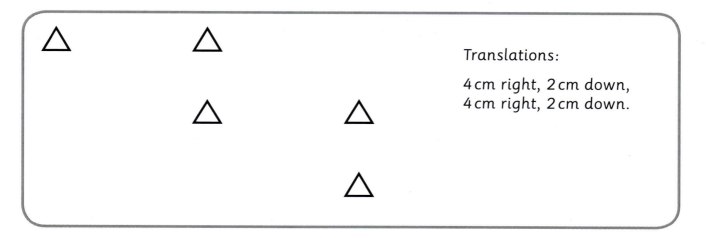

Translations:

4 cm right, 2 cm down,
4 cm right, 2 cm down.

Angles 1

- Multiply by 25 by multiplying by 100 and dividing by 4. (5Nc14)
- Use factors to multiply, e.g. multiply by 3, then double to multiply by 6. (5Nc15)
- Understand and use angle measure in degrees; measure angles to the nearest 5°; identify, describe and estimate the size of angles and classify them as acute, right or obtuse. (5Gs6)

Pencils; paper; photocopiable page 173; rulers.

Starter

- Tell the learners they will be rehearsing mental calculation strategies for multiplication.
- Call out multiples of 4. The learners multiply them by 25 by multiplying by 100 and dividing by 4. They use jottings to help them find the answer.
- Now call out other numbers. The learners multiply these by 3 then double to multiply by 6. They also multiply by 4 then double to multiply by 8.

Main activities

- Ask: *What is an angle?* Listen to the learners' responses. Establish that there are two types of angle. One is a dynamic angle which is to do with direction of turn, for example, compass directions. The second is a static angle which is the amount of turn between two lines that meet at a point.
- Tell the learners they will focus on static angles. Draw this on the board:

Explain that the curve shows the angle between the two lines.

- Discuss these types of angles: right (perpendicular lines), acute (smaller than a right angle) and obtuse (larger than a right angle, smaller than a straight line). Ask the learners to make all three types of angle using their arms.
- Ask the learners to draw a selection of right, acute and obtuse angles and pass their drawings to a partner to identify.
- Now ask the learners to draw a variety of polygons and mark the acute angles with a dot and the obtuse angles with a cross.
- Ask the learners to work through photocopiable page 173.

Plenary

- Take feedback from the photocopiable activity. Invite individuals to share their angles and their shapes. Ask the class to identify both.
- Recap what dynamic and static angles are.

Ask the learners:

- What is a dynamic angle? Can you explain in another way? Can you give an example?
- How would you describe a static angle? Is there another way?
- What word can we use to describe the size of static angles? Can you tell me another?

Support: At the end of the photocopiable activity, ask these learners to draw three different shapes, one with a right angle, one with an acute angle and one with an obtuse angle.

Extension: After they have completed the photocopiable activity, ask these learners to measure the angles they drew using a protractor.

Name: _____

Angles

You will need:

A ruler and a pencil.

What to do

- In the spaces below, draw at least two more examples of right, acute and obtuse angles.

Right angles

Acute angles

Obtuse angles

- Now draw two shapes that have at least one right angle, one acute angle and one obtuse angle inside them. Name your shapes.

  Here is an example:

  Heptagon

Angles 2

Learning objectives

- Understand and use angle measure in degrees; measure angles to the nearest 5°; identify, describe and estimate the size of angles and classify them as acute, right or obtuse. (5Gs6)
- Calculate angles in a straight line. (5Gs7)

Resources

Pencils; paper for recording; protractors; A4 paper; rulers; photocopiable page 175.

Starter

- Tell the learners they will practise multiplying by 25.
- Call out any two-digit numbers, for example: *34, 15*. The learners multiply them by 25 by multiplying by 100 and dividing by 4 (34: 3400, 1700, 850). They use jottings to help them find the answer.
- Repeat for simple three-digit numbers.

Main activities

- Ask: *What can you tell me about angles? Talk to your partner.* Take feedback. Agree that there are two types of angle (dynamic and static) and that angles can be right, acute and obtuse. Ask them to draw some of each of these.
- Ask: *Does anyone know how we measure the size of angles?* Establish that they are measured in units called degrees. These units are different from those used for temperature. Tell them that a right angle measures 90°. Ask them to use this information to work out the angle in a straight line (two right angles) and a circle (four right angles).
- Give each learner a protractor. Ask them to look at the scale. Tell them that it is basically a number line. Call out various multiples of 10° for them to find, for example: *60°, 120°*. Repeat for multiples of 5°.

- Give each learner a piece of A4 paper, ruler and pencil. Ask then to draw a line. Ask them to look at their protractor and find the line parallel to its straight edge. Tell them to put this line on the line they have drawn. Say: *55°*. The learners should make a mark on their paper beside 55. Repeat for other angles.
- Draw two lines that meet on the board. Demonstrate how to use a protractor to measure the angle.
- Ask the learners to work through photocopiable page 175.

Plenary

- Take feedback from the photocopiable activity. Invite individuals to share their estimates and the actual sizes of the angles. Ask volunteers to draw their triangle on the board and to explain how they found the angles at the two corners.
- Discuss where angles are important in real life, for example building, map reading.

Success criteria

Ask the learners:

- What can you tell me about angles? Is there anything else?
- How would you explain how to use a protractor to someone who didn't know? Is there another way?
- Who would use angles in their work or hobby? Can you think of anyone else?

Ideas for differentiation

Support: During the photocopiable activity, ask these learners to measure to the nearest 5°. Ask them to draw a right-angled isosceles triangle and find one angle.

Extension: During the photocopiable activity, challenge these learners to measure their angles to the nearest degree.

Measuring angles

You will need:

A pencil and a protractor.

What to do

- Use your knowledge that a right angle is 90° to estimate the sizes of these angles.

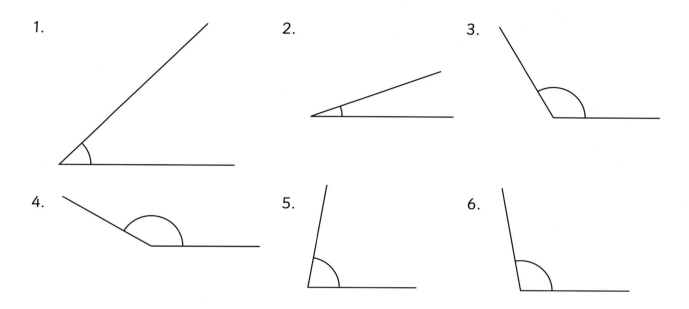

1.   2.   3.
4.   5.   6.

- Write your estimates in the table below.

Angle	Estimate	Actual	Difference
1			
2			
3			
4			
5			
6			

- When you have estimated each one, check using a protractor.
- Add the information to the table.
- You can then use the actual sizes to help you make your next estimate.
- Draw a right-angled scalene triangle on the back of this paper.
- What are the sizes of the other two angles? Label them inside your triangle.

Unit assessment

Questions to ask

- Why can't a triangle have two right angles?
- Where would you see perpendicular lines?
- Describe what the net of a triangular prism would look like.

- What is a rotation?
- How would you measure an angle?

Summative assessment activities

Observe the learners while they take part in these activities. You will quickly be able to identify those who appear to be confident and those who may need additional support.

Angles

This activity assesses the learners' understanding of measuring angles.

You will need:

Paper; pencils; rulers; protractors; six angles drawn on paper (as on page 175) for each learner.

What to do

- Organise the learners into groups of four.
- Give each learner a copy of the angles you have drawn.
- Ask them to consider the type of angle each one is (acute, right or obtuse) and label them accordingly.
- Then ask them to estimate the size of each angle and write their estimates on a piece of paper.
- Ask them to measure the angles using a protractor and compare their estimates with the actual measurements.

Which shape?

This game assesses the learners' knowledge of 2D shapes.

You will need:

Paper; pencils.

What to do

- Organise the learners into groups of three or four.
- Describe a variety of 2D shapes to the learners. For example: *This shape has three sides. One angle is a right angle and the other two are equal.*
- The learners should work out which shape you are describing and draw it on a piece of paper.

Written assessment

Distribute copies of photocopiable page 177. Ask the learners to read the questions and write the answers. They should work independently.

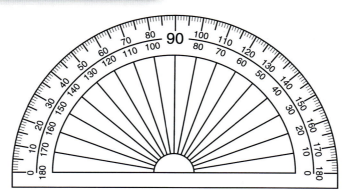

Name: _____

Working with shapes 2

1. Draw a right-angled isosceles triangle.

2. Draw a scalene triangle with an obtuse angle.

3. Draw a shape with one pair of parallel sides.

4. Draw a shape with two pairs of perpendicular sides.

5. Draw an angle of your choice. Measure it and write this beside your angle.

6. Rotate this arrow 180° to the right:

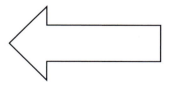

7. Reflect this shape across the diagonal mirror line:

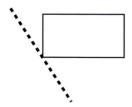

8. Translate this shape 3 centimetres to the right and 2 centimetres down:

Measures 1

Learning objectives

● Read, choose, use and record standard units to estimate and measure length, mass and capacity to a suitable degree of accuracy. (5MI1)

● Convert larger to smaller metric units (decimals to one place), e.g. change 2.6 kg to 2600 g. (5MI2)

● Draw and measure lines to the nearest centimetre and millimetre. (5MI7)

Resources

Plain paper; pencils; rulers; photocopiable page 179; sand; weighing scales; books; plastic bags; 2-litre water bottles; measuring jugs or cylinders.

Starter

- Explain that the learners will practise converting and ordering units of measure to help in the lesson.
- Call out different amounts relating to length. Ask the learners to write these amounts in at least two different ways, for example you say: *254 cm*, they write 2 m 54 cm, 2.54 m.
- When you have said a selection of different lengths, the learners order these from smallest to largest.
- Repeat for mass and capacity.

Main activities

- Ask: *What do you know about measurement? Discuss this with a partner.* Take feedback. Expect them to talk about length, mass and capacity. They should tell you what each is a measurement of, the different units used and the measuring equipment associated with each.
- Ask the learners to give examples of what measuring instruments could be used to measure:
  - the height of the table: ruler, metre stick, tape measure
  - the mass of a book: balances, kitchen scales
  - the capacity of a mug: measuring jug.

- Focus on the units of length. Ask the learners to tell you those they know, both metric and imperial if appropriate. As they tell you, ask for their abbreviations. Write these on the board. Discuss equivalences and ask them to tell you, for example, how many millimetres in 3.4 cm, metres in 12.5 km.
- Call out some centimetre and millimetre lengths for them to draw accurately on a piece of plain paper, for example: *3.4 cm, 145 mm*. Once they have finished they should swap paper with a partner and check each other's measuring.
- Distribute photocopiable page 179.

Plenary

- Invite groups to share what they did during the photocopiable activity. Did they agree with the statements? How did they come to their decision?
- Recap equivalent units of length, mass and capacity by asking the learners to tell you the different ways to express such measurements as 1.3 km, 2.6 kg, 1.9 litres.

Success criteria

Ask the learners:

● What units do we use to measure length / mass / capacity? Are there any others?

● What measuring equipment can we use to measure length / mass / capacity? Are there any others?

● How many millimetres are there in 12.4 cm? How do you know?

● Which is longer, 1.05 km or 1500 m? How do you know?

Ideas for differentiation

Support: During the photocopiable activity, ask these learners to work in a mixed attaining group. This will facilitate peer support.

Extension: Once these learners have completed the photocopiable activity, ask them to make up their own statement to investigate.

Measures

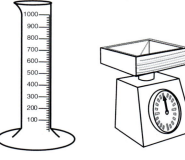

You will need:

Sand, weighing scales, a book, plastic bags, a 2-litre water bottle and a measuring jug or cylinder.

What to do

- Work in a group of four.

- Decide which activity you will do first. Read the statement and decide how you can find out if it is true or not.

- Collect the equipment you need.

- Try out your ideas. Write down what you did and what you found out.

- Once you have done this, try the other activity.

Activity 1

Pierre says:

I can make three different masses using bags of sand. These will help me estimate the mass of a dictionary.

a) What do you think?

b) How are you going to find out?

c) Do you agree with Pierre?

Activity 2

Paddy says:

I think that a 2-litre water bottle really has a capacity of a little more than 2 litres.

a) What do you think?

b) How are you going to find out?

c) Do you agree with Paddy?

Measures 2

Learning objectives

- Read, choose, use and record standard units to estimate and measure length, mass and capacity to a suitable degree of accuracy. (5MI1)
- Order measurements in mixed units. (5MI3)
- Round measurements to the nearest whole unit. (5MI4)
- Compare readings on different scales. (5MI6)

Resources

Pencils; paper; 1 kg bag of sand (or similar); weighing scales; loose sand; plastic bags; photocopiable page 181; two measuring cylinders with different scales; water; 1- and 2-litre bottles; selection of containers; measuring jugs.

Starter

- Explain that the learners will practise ordering and rounding units of measure.
- Write several decimal lengths on the board, for example: *1.4 cm, 2.54 m, 12.5 m, 1.246 km.* Ask the learners to write these in order from smallest to largest, then round them to the nearest whole unit.
- Repeat for units of mass and capacity.

Main activities

- Recap the vocabulary of length, for example 'height', 'width'. Ask: *What would we measure in millimetres?* Repeat this question for centimetres, metres and kilometres. Ask: *Can you think of any other units of length?* They may say, for example, miles, feet, inches. If appropriate, compare the metric and imperial units, for example three feet is approximately one metre.
- Repeat for mass and capacity.
- Show the 1 kg bag of sand. Invite volunteers to feel its mass. Then ask the learners to find items in the classroom that are lighter, heavier and approximately the same. They should check their examples on a set of scales. Invite an individual to describe how to read the scale.

- Ask the learners to work in groups of four. Each group needs sand, two plastic bags and weighing scales. Ask them to put the sand into the bags so that one is heavier than the other. Then ask them to estimate and weigh the two masses.
- Ask the learners to use their bags of sand to estimate the weight of different items in the classroom. They should come to an agreed estimate for each item and then weigh it to see how close their estimates were.
- Ask the learners to work through photocopiable page 181.

Plenary

- Take feedback from the photocopiable activity. Invite groups to share with the class the containers they selected, their estimates and the actual capacities.
- Invite a group to pour the water from one of their containers into two different measuring cylinders and compare the appearance and how the scales differ.

Success criteria

Ask the learners:

- How can we make a sensible estimate of the mass of an item? How else?
- What words tell us about length? Can you think of any others?
- What words tell us about mass? Can you think of any others?
- What would we measure in metres? What about kilograms? Millilitres?

Ideas for differentiation

Support: During the photocopiable activity, ask these learners to work in a mixed attaining group. This will facilitate peer support.

Extension: Once they have completed the photocopiable activity, ask these learners to estimate and measure the capacity of some small containers to the nearest millilitre.

Name: _____

Estimating capacity

You will need:

Water, 1- and 2-litre bottles of water, four different containers and a measuring jug.

What to do

- Work in a group of four.

- Collect four different containers from around the classroom. They all need to look different.

- As a group, estimate the capacity of one of your containers. Write your estimate in the table below.

- Measure the amount you estimated into a measuring jug and see if it fills the container. If your estimate was not correct, find out the actual capacity of the container. Add this information to the table.

- Repeat this for the other three containers.

Estimate	Was your estimate correct?	Actual capacity

Time 1

Learning objectives

- Interpret a reading that lies between two unnumbered divisions on a scale. (5MI5)
- Tell and compare the time using digital and analogue clocks using the 24-hour clock. (5Mt2)
- Understand everyday systems of measurement in length, weight, capacity, temperature and time and use these to perform simple calculations. (5Pt1)
- Consider whether an answer is reasonable in the context of a problem. (5Pt7)
- Deduce new information from existing information to solve problems. (5Ps4)

Resources

Counting stick; pencils; paper; individual analogue clock faces; photocopiable page 183.

Starter

- Explain that the learners will rehearse reading numbers that are on an unmarked scale.
- Show the counting stick. Tell the learners zero is at one end and 1 metre is at the other.
- Point to divisions and also to points between divisions. Ask the learners to tell you what measurement belongs at each position.
- Repeat for measurements from 0 to 1 kg, 0 to 1 litre, 2 m to 3 m and so on.

Main activities

- Give the learners a clock face each. Ask them to show you different times and write down the equivalent 24-hour digital time. For example, say: *5 minutes to 9 in the evening.* They should show that time and write 20:55.
- Set problems such as: *The film started at 17:45. Bobby was 35 minutes early. At what time did he arrive at the movie theatre / cinema?* Ask the learners to show the time he arrived on their clock faces. Encourage them to check their answer to see if it is reasonable.

- Ask: *What is a really helpful tool to use to work out time differences?* Agree on a number line. Invite volunteers to demonstrate how to use one to solve the problems you asked, for example:

See page 122 for more details.

- Give some departure and arrival times of a bus, train or plane. Ask the learners to use number lines to work out the journey times, for example: *The bus departs at 09:15 and arrives at its destination at 15:40. How long did the journey take?*
- Ask the learners to work through photocopiable page 183.

Plenary

- Invite the learners to share the problems they made up in the photocopiable activity. The class answers them.
- Ask the learners to assess their confidence in finding time differences using a number line.

Success criteria

Ask the learners:

- What is another way of saying 25 minutes to 11? Is there another way?
- If a bus leaves the station at 13:12 and arrives back nine and a half hours later, what time does it return? How did you work that out? Is there another way?
- Sammy walked for 1 hour and 8 minutes. If he left home at 14:30, when did he get back?

Ideas for differentiation

Support: Adapt photocopiable page 183 so that these learners work out differences between whole, half and quarter hour intervals.

Extension: Adapt photocopiable page 183 so that these learners are working out differences between minute intervals.

Name: _____

More problems!

1. Solve these problems. Draw number lines to help you.
    Each time check your answer to see if the answer is
    reasonable.

    a) Cherri went strawberry picking. She began
        at 10:20 and was picking strawberries for
        2 hours 45 minutes.

        When did she finish?

        My answer: _____

    b) Anan spent 1 hour 55 minutes at the gym. She left at 16:30.

        When did she get there?

        My answer: _____

    c) The twins went to the beach. They arrived at 11:50 and left at 17:15.

        How long were they at the beach for?

        My answer: _____

    d) Zeina and Mona left for school at 07:15. They spent the day working hard.
        They got home at 17:05.

        How long were they away from home?

        My answer: _____

    e) Charlie and Chris were gardening. They started at 13:25. Charlie finished
        at 15:55. Chris carried on for another hour and ten minutes.

        For how long was Chris gardening?

        My answer: _____

2. Now make up and solve some problems of your own on the back of this paper.

Time 2

Learning objectives

- Recognise and use the units for time (seconds, minutes, hours, days, months and years). (5Mt1)
- Read timetables using the 24-hour clock. (5Mt3)
- Calculate time intervals in seconds, minutes and hours using digital or analogue formats. (5Mt4)
- Use ordered lists and tables to help solve problems systematically. (5Ps5)

Resources

Set of cards from photocopiable page 121; pencils; paper; photocopiable page 185.

Starter

- As a class play the 'just a minute' game from photocopiable page 121. Give the meanings (such as 24 hours) and ask the learners to guess the words (a day). Do this for one minute and then count how many they identified correctly.
- Repeat: do they get faster?

Main activities

- Ask: *What can you tell me about time?* Encourage them to talk about digital and analogue clocks and the vocabulary of time, for example 'seconds', 'minutes', 'weeks', 'months'.
- Ask them to give you statements related to the units of time, for example, there are 180 minutes in 3 hours, 120 seconds in two minutes.
- Ask: *When do we usually see 24-hour clock time displayed?* Agree on such items as computers, microwave ovens and also timetables. Ask the learners to tell you what a timetable is, for example, they give information about when buses / trains / planes arrive and depart. They also give information about television programmes and showing times for films at the cinema / movie theatre.

- Together make up a TV guide. Draw a table on the board. Ask the learners to suggest TV programmes. Add these to the table. Make up start and finish times and add these. Ensure they last for a variety of amounts of time, for example, 23 minutes, one and a half hours.
- Ask questions such as: *How long is X on for? Which is the longest TV programme?* Invite the learners to ask each other questions from the table.
- Ask the learners to work through photocopiable page 185.

Plenary

- Take feedback from the photocopiable activity, inviting the learners to share their timetables and give information from it. Invite the rest of the class to ask them questions.
- Ask the learners to assess their confidence in working out information from timetables.

Success criteria

Ask the learners:

- What is 18:24 as a 12-hour clock time? How did you work that out? Is there another way?
- What is the time difference between 12:36 and 17:05? How did you work that out? Is there another way?
- If the television programme lasted for 1 hour 45 minutes and finished at 18:30, what time did it begin? How did you work that out?

Ideas for differentiation

Support: During the photocopiable activity, ask these learners to make up times that are whole, half and quarter hours in length.

Extension: During the photocopiable activity, ask these learners to make up times that show different numbers of minutes, for example 17:23.

Name: _____

TV guide

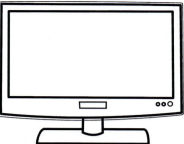

1. Take a survey of the favourite television programmes of the class. Pick the six most popular.

2. Make up a TV guide for these programmes.

3. Make up the start time and then the finish time. Each programme should follow on from the previous one. Each programme needs to last a different amount of time from the others. Work out how long each programme lasts.

4. Put all this information in the table below. Two examples have been done for you.

TV programme	Start time	Finish time	Length of programme
The Pirate's Treasure Hunt	16:00	16:25	25 minutes
The News	16:25	16:40	15 minutes

Area and perimeter 1

Starter

- Tell the learners they will rehearse multiplication and division facts.
- Share out the follow-me cards between small groups. Keep one card. Read out the multiplication, for example: *4 × 6*. The group with the answer calls it out, then reads the multiplication on their card and so on.
- Repeat, starting with the answer on your card for division.

Main activities

- Ask: *What do we mean by perimeter and area? Talk to your partner.* Take feedback. Agree that area is the amount of space taken up, perimeter is the outside of an area.
- Ask the learners to give examples of where they would find perimeters and areas.
- Recap the units used to measure perimeter (different lengths) and area (squares of those lengths).
- Ask: *How can we find the perimeter of a rectangle?* Agree that you can add the lengths of the sides or use the formula $2(l + w)$.

- Ask: *How can we find the area of a rectangle?* Agree that you multiply length by width, using the formula $l \times w$.
- Set this problem: *Ahmed is going to sow grass seed in his garden and build a wall around it. It is a rectangle measuring 8 m by 4.5 m. He needs to know the perimeter and area so he can buy the grass seed and bricks for the wall.* Discuss how the learners can find the perimeter and area. Ask the learners to work them out. Agree that the perimeter is $2(8\,m + 4.5\,m) = 25\,m$, and the area is $8\,m \times 4.5\,m = 36$ square metres.
- Repeat with similar problems.
- Ask the learners to work through photocopiable page 187.

Plenary

- Take feedback from the photocopiable activity. Together find all the possible whole number perimeters that can be found for Sophie's area.
- Discuss which is probably the best shape for her patio and why.

Name: _____

Different areas

You will need:

A ruler and a pencil.

What to do

- Sophie would like to build a rectangular patio in her garden.
  She wants the area of her patio to be 24 square metres.

- Think about the possible sizes that Sophie's patio could be. Write these down.

- Draw some designs using these sizes. Use a scale of 1 cm = 1 m. Measure
  accurately using your ruler. Label the measurements.

- Once you have drawn your rectangles, check to make sure the areas are correct.

- Write down the perimeter of each shape.

  Here is an example:

  2 m ⌐─────────────────────────────⌐   Perimeter = 28 m
      | |
      └─────────────────────────────┘
                 12 m

- Which one of your designs would you advise Sophie to use? Explain why on the back
  of this paper.

Area and perimeter 2

Learning objectives

● Use a calendar to calculate time intervals in days and weeks (using knowledge of days in calendar months). (5Mt5)

● Calculate time intervals in months or years. (5Mt6)

● Measure and calculate the perimeter of regular and irregular polygons. (5Ma1)

● Investigate a simple general statement by finding examples which do or do not satisfy it, e.g. the sum of three consecutive whole numbers is always a multiple of three. (5Ps8)

Resources

Photocopiable page 192; centimetre squared paper; pencils; rulers; photocopiable page 189.

Starter

- Explain to the learners that they will practise using a calendar.
- Give pairs of learners a copy of photocopiable page 192. Ask questions relating to time intervals in days and weeks and also months, for example:
  - *How long is it from the fourth Tuesday in May to the third Wednesday in September?*
  - *What date will it be five months, three weeks and two days after 2nd March?*
  - *What date will it be in one year and three months' time?*

Main activities

- Ask: *What formula can we use to find the perimeter of a rectangle?* Agree on 2(*l* + *w*) or 2*l* + 2*w*.
- Ask: *What formula can we use to find the area of a rectangle?* Agree on *l* × *w*.
- Ask the learners to draw some rectangles on a piece of squared paper. They choose the lengths of the shapes' sides. They should work out their perimeters and areas using the formulae discussed.
- Ask: *If the perimeter gets bigger, so does the area. Is this statement true?* Agree that it isn't always true. Ask the learners to prove it by drawing rectangles with different perimeters but the same area on their squared paper.

- Ask: *If the area gets bigger, so does the perimeter. Is this statement true?* Agree that this isn't always true. Set this problem so that they can prove it: *Ali has 36 m of fencing for a pen for his sheep. What different areas can he make?* Ask the learners to work with a partner. They should draw possible shapes on squared paper and work out the areas, for example 2 m × 16 m: area = 32 square metres, 4 m × 14 m: area = 56 square metres.
- Ask the learners to work through photocopiable page 189.

Plenary

- Invite the learners to share the different areas they found for the 50 m perimeter in the photocopiable activity.
- Discuss the best perimeter for the stable, including reasons why.
- Recap that area and perimeter don't have a relationship with each other.

Success criteria

Ask the learners:

● Is this statement always true: if the perimeter gets bigger so does the area? Why not?

● Is this statement always true: if the area gets bigger so does the perimeter? Why not?

● Can you give an example of when the perimeter stays the same and the area is different? Can you give another?

Ideas for differentiation

Support: During all the activities, allow these learners to draw their shapes on squared paper.

Extension: When they have completed the photocopiable activity, tell these learners that the material Kamrun is going to use for the floor of the stable costs $120 per square metre. Ask them to work out the cost for each area.

Name: _____

Different perimeters

You will need:

A ruler and a pencil.

What to do

- Kamrun has been given a large area of land. He would like to build a stable for his horse on part of it. He wants the stable to be rectangular with a perimeter of 50 m.

- Use the space below to work out some of the possible areas for Kamrun's stable. Write these down.

- Sketch some designs using these sizes. Your measurements to do not need to be to scale. Remember to label them.

- Once you have drawn your rectangles, check to make sure the perimeters are correct.

- Write down the area of each shape.

  Here is an example:

  5 m | Area = 100 square metres

  20 m

- Which one of your designs would you advise Kamrun to use? Explain why on the back of this paper.

Unit assessment

Questions to ask

- Tell me some equivalent lengths, for example 1.4 km = 1 km 400 m = 1400 m.
- Which units do we use to measure capacity?
- What measuring equipment can we use to measure mass?

- Is it true that as the perimeter gets longer the area gets larger?
- What is meant by area?

Summative assessment activities

Observe the learners while they take part in these activities. You will quickly be able to identify those who appear to be confident and those who may need additional support.

What's the time?

This activity assesses the learners' understanding of 24-hour time.

You will need:

Paper; pencils; individual clock faces.

What to do

- Organise the learners into groups of four.
- Give each learner a clock face.
- Call out various 12-hour times, for example 6:40 a.m.
- Ask the learners to show these times on their clock face and write them in 24-hour time.

Timetables

This activity assesses the learners' knowledge of finding time differences.

You will need:

Pencils; paper.

What to do

- Organise the learners into groups of four.

- Ask questions such as:
  - *The first train leaves the depot at 07:35 and the next one leaves 32 minutes later. At what time does it leave?*
  - *Train 3 leaves at 08:26. How much later does this leave than train 2?*
  - *Train 4 leaves the depot at 12:27. Train 5 leaves 1 hour and 40 minutes later. What time is that?*
- Encourage them to use a time number line.

Estimating and measuring length

This game assesses the learners' ability to measure accurately.

You will need:

Paper; pencils; four pieces of string of different lengths; rulers; tape measures.

What to do

- Organise the learners into groups of three or four.
- Ask the learners to look at each piece of string in turn. They should first estimate each length and write this down on paper.
- They then measure each as accurately as they can.
- They compare their estimates with the actual measurement to see how close they were.

Written assessment

Distribute copies of photocopiable page 191. Ask the learners to read the questions and write the answers. They should work independently.

Name: _____

Working with measures 3

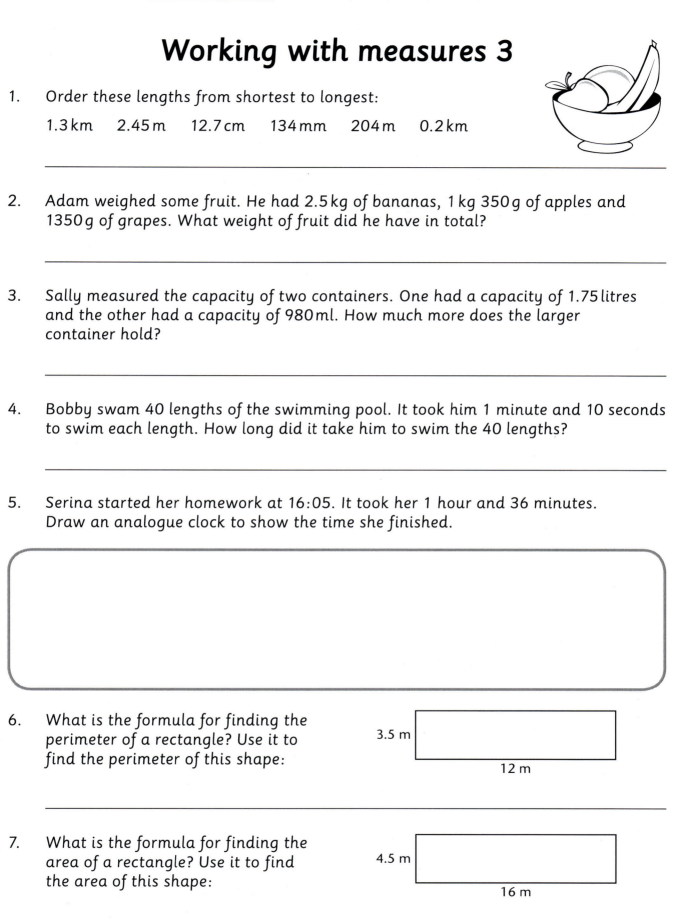

1. Order these lengths from shortest to longest:

    1.3 km   2.45 m   12.7 cm   134 mm   204 m   0.2 km

    _____

2. Adam weighed some fruit. He had 2.5 kg of bananas, 1 kg 350 g of apples and
    1350 g of grapes. What weight of fruit did he have in total?

    _____

3. Sally measured the capacity of two containers. One had a capacity of 1.75 litres
    and the other had a capacity of 980 ml. How much more does the larger
    container hold?

    _____

4. Bobby swam 40 lengths of the swimming pool. It took him 1 minute and 10 seconds
    to swim each length. How long did it take him to swim the 40 lengths?

    _____

5. Serina started her homework at 16:05. It took her 1 hour and 36 minutes.
    Draw an analogue clock to show the time she finished.

6. What is the formula for finding the
    perimeter of a rectangle? Use it to
    find the perimeter of this shape:

    3.5 m

    12 m

    _____

7. What is the formula for finding the
    area of a rectangle? Use it to find
    the area of this shape:

    4.5 m

    16 m

    _____

Calendar

2012

January

S	M	T	W	T	F	S
1	2	3	4	5	6	7
8	9	10	11	12	13	14
15	16	17	18	19	20	21
22	23	24	25	26	27	28
29	30	31				

February

S	M	T	W	T	F	S
			1	2	3	4
5	6	7	8	9	10	11
12	13	14	15	16	17	18
19	20	21	22	23	24	25
26	27	28	29			

March

S	M	T	W	T	F	S
				1	2	3
4	5	6	7	8	9	10
11	12	13	14	15	16	17
18	19	20	21	22	23	24
25	26	27	28	29	30	31

April

S	M	T	W	T	F	S
1	2	3	4	5	6	7
8	9	10	11	12	13	14
15	16	17	18	19	20	21
22	23	24	25	26	27	28
29	30					

May

S	M	T	W	T	F	S
		1	2	3	4	5
6	7	8	9	10	11	12
13	14	15	16	17	18	19
20	21	22	23	24	25	26
27	28	29	30	31		

June

S	M	T	W	T	F	S
					1	2
3	4	5	6	7	8	9
10	11	12	13	14	15	16
17	18	19	20	21	22	23
24	25	26	27	28	29	30

July

S	M	T	W	T	F	S
1	2	3	4	5	6	7
8	9	10	11	12	13	14
15	16	17	18	19	20	21
22	23	24	25	26	27	28
29	30	31				

August

S	M	T	W	T	F	S
			1	2	3	4
5	6	7	8	9	10	11
12	13	14	15	16	17	18
19	20	21	22	23	24	25
26	27	28	29	30	31	

September

S	M	T	W	T	F	S
						1
2	3	4	5	6	7	8
9	10	11	12	13	14	15
16	17	18	19	20	21	22
23	24	25	26	27	28	29
30						

October

S	M	T	W	T	F	S
	1	2	3	4	5	6
7	8	9	10	11	12	13
14	15	16	17	18	19	20
21	22	23	24	25	26	27
28	29	30	31			

November

S	M	T	W	T	F	S
				1	2	3
4	5	6	7	8	9	10
11	12	13	14	15	16	17
18	19	20	21	22	23	24
25	26	27	28	29	30	

December

S	M	T	W	T	F	S
						1
2	3	4	5	6	7	8
9	10	11	12	13	14	15
16	17	18	19	20	21	22
23	24	25	26	27	28	29
30	31					

Cambridge Primary: Ready to Go Lessons for Maths Stage 5 © Hodder & Stoughton Ltd 2012